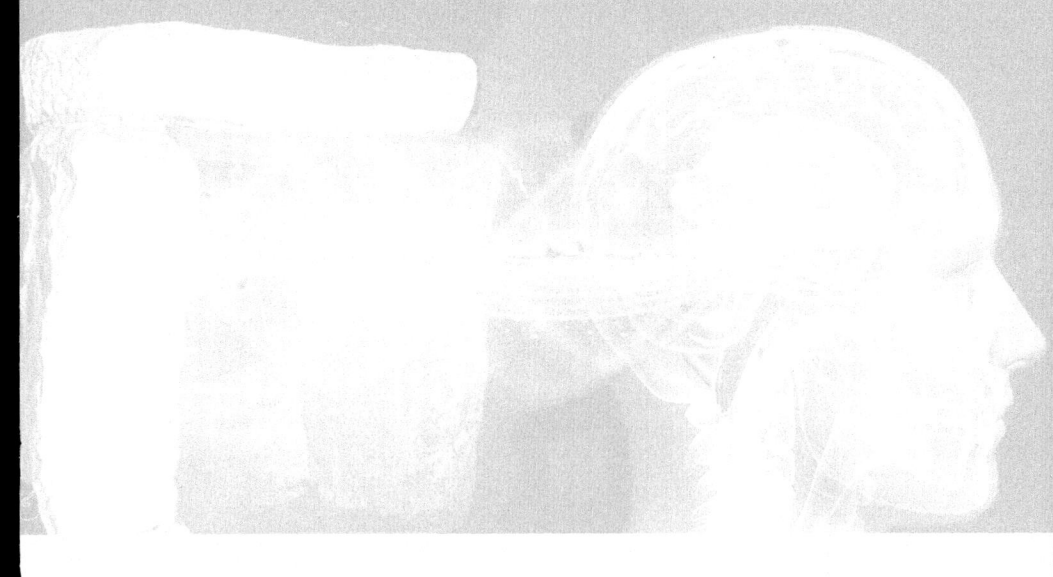

MEGALITHS, MUSIC, AND THE MIND

MEGALITHS, MUSIC, AND THE MIND

Linda C. Eneix

Jenny Stanford
Publishing

Published by

Jenny Stanford Publishing Pte. Ltd.
101 Thomson Road
#06-01, United Square
Singapore 307591

Email: editorial@jennystanford.com
Web: www.jennystanford.com

British Library Cataloguing-in-Publication Data
A catalogue record for this book is available from the British Library.

Megaliths, Music, and the Mind
Copyright © 2024 Jenny Stanford Publishing Pte. Ltd.

All rights reserved. This book, or parts thereof, may not be reproduced in any form or by any means, electronic or mechanical, including photocopying, recording or any information storage and retrieval system now known or to be invented, without written permission from the publisher.

For photocopying of material in this volume, please pay a copying fee through the Copyright Clearance Center, Inc., 222 Rosewood Drive, Danvers, MA 01923, USA. In this case permission to photocopy is not required from the publisher.

ISBN 978-981-5129-25-0 (Hardcover)
ISBN 978-1-003-46823-3 (eBook)

Contents

	Preface	vii
1.	**Introduction**	**1**
	1.1 The "Sapient Paradox" Question	4
	1.2 Emergence of a New Field of Study	11
	1.3 Assembling Expert Voices	12
2.	**The Ħal Saflieni Project**	**21**
	2.1 The Megalithic Range	35
	2.2 Human Response	40
3.	**Orientation**	**51**
	3.1 The Sapient Development Timeline	51
	3.2 Neolithic Revolution	58
	3.3 Genetics and Migrations	62
4.	**Megaliths**	**73**
	4.1 The Earliest	73
	4.2 Placement	83
	4.3 Design	85
	4.4 Acoustification/Intention	96
	4.5 Materials	97
	4.6 Incentive and Project Management	100
5.	**Music**	**103**
	5.1 Instruments	106
	5.2 Melody	112

6. The Mind — 119

    6.1   The Spiritual Element — 119
    6.2   Magic? — 126
    6.3   Art — 130
    6.4   Neuroscience — 147

7. The Present of Archaeoacoustics — 159

    7.1   Pseudoscience and the Wall of Resistance — 159
    7.2   Research — 161
    7.3   Application — 164

References — 169

BONUS MATERIAL

Resonant Form: The Convergence of Sound and Space — 177

The Ħal Saflieni Hypogeum: A Link between Palaeolithic Painted Caves and Romanesque Chapels? — 195

Fear and Amazement — 209

Sound, Cognition, and Social Control — 231

Voicing Cave: Experience and Metaphor, from Archaeoacoustics to Voice Therapy — 241

Initiation: Inside the Great Pyramid — 257

Index — 261

Preface

A first encounter with the megalithic monuments on the Mediterranean islands of Malta changed my life in 1990. No stranger to antiquity, I had already climbed inside the great pyramid at Giza and walked the streets of the Old City in Jerusalem by that time. The authenticity of ancient stone could be recognized, or it would have been easy to think that the Maltese temples were something modern, created to draw tourists. But no. They were the real deal and they were (at that time) older than anything on the planet made by humankind. The desire to know more about the mysterious advanced society that built them was irresistible. I returned home to the USA and searched for books on the subject, but could find none. In fact, there was a glaring hole in the books about the development of early civilizations that *could* be found. Somehow, over the course of many visits to Malta, it became some sort of cosmic assignment for me to collect every shred of data. Thus, it was with some prior knowledge about the Hal Saflieni Hypogeum and a growing suspicion that I encountered the subject of archaeoacoustics. Then came the first reports from Göbekli Tepe and a subsequent visit to Dr. Klaus Schmidt.

As elaborated in this book, the past was not silent and our ancestors were not deaf. Yet the element of sound has been largely absent in our consideration of the ancient world. I suppose it had to be someone like me who was not academically confined to one narrow field, but had the ability and interest to start collecting reports from a wide range to address the question of: "What has sound got to do with the building of the first megalithic monuments and the subsequent lifestyle changes that launched Western Civilization?" The answer is here and it is a revelation.

Megaliths, Music, and the Mind assembles content from the worlds of archaeology, architecture, anthropology, ethnomusicology, genetics, neuroscience, physics, and more. Fascinating pieces of evidence are set side by side, resulting in a stunning premise. The book is a core overview focused on the rediscovery of an ill-understood sensory element of developing culture, with hope for therapeutic application in the modern world. As the pieces come together, this exploration of the human experience of special sound in ancient ritual and ceremonial spaces brings new perspective for anyone with an interest in prehistory and human development in its most pivotal days. Hopefully it will also stimulate thought and suggest areas of expansion for ongoing researchers.

Chapter 1

Introduction

In the pages that follow, we are exploring the advent of Western civilization with our ears open.

Exercise

Before you begin reading, please take just a few minutes to consciously listen to your environment and mentally catalog what you are hearing right this minute. An air-handler, perhaps? Traffic outside? Something that lets you know whether there are other people around and what they might be doing? Or are you hearing complete silence? How about your own breathing?

We are bringing focus to what is a natural part of living for most people, although we filter differently these days. The human capacity for listening was functioning just as well thousands of years ago as it does today. The past was not silent.

The reader will be seeing more of these boxes which suggest possible group activity in a classroom, around a campfire or when the conversation dies around a table.

Megaliths, Music, and the Mind
Linda C. Eneix
Copyright © 2024 Jenny Stanford Publishing Pte. Ltd.
ISBN 978-981-5129-25-0 (Hardcover), 978-1-003-46823-3 (eBook)
www.jennystanford.com

A close examination of human development before history (and without which history would not exist) is imperative if we are going to confront complex and interrelated problems facing us today. Our modern habits, traditions, values and morals are the result of events and philosophy that have been handed down to us since history began with the invention of writing. When we started down the road of "civilization", we couldn't know where it was going to take us. Now, Western technological development is impacting every corner of the planet. As we begin to question how much of this escalated development is sustainable in the face of dwindling natural resources and increasing challenges with its influence on mental health, it is certainly worth trying to understand all the aspects of how we came to be who we are.

If you had to start from scratch, could you make a cell phone? Could you make a laptop or a television? Could you make these devices work? Of course not. There is a tremendous infrastructure of technology and knowledge that forms the foundation for these miraculous gadgets and the resources that make them function. It comes from people who have been building on the knowledge of those who came before them. The ability to improve and expand on the invention of others to this extent largely stems from what we call civilization. Rather than a literal "living in cities" definition, we refer here to societies that have organized their administration, learned how to control their food supply and managed to settle for an extended period more or less in one place. The very first such societies appeared at the end of the Stone Age, which is about much more than using stone tools, which is the full definition given in many search engines. The world as we know it would never be possible without the lifestyle that was born in the Neolithic era, or New Stone Age, as the transition period is called. How did that happen? Can we learn anything useful from it to apply to life today?

Figure 1.1 Mnajdra Temple Complex (South temple), Malta, ca. 5,600 years old.

1.1 The "Sapient Paradox" Question

The explanation for a radical change in lifestyle that has shaped ensuing human development has remained baffling. The Big Question that researchers of prehistory ask is essentially: why did humankind, after 200,000 years of nomadic hunting and gathering, begin developing agriculture which subsequently resulted in the settled lifestyle that sent us into Western civilization? With the discovery and dating of the Göbekli Tepe archaeological site in the 1990s, we now know that agriculture followed closely on the heels of the building of the first great stone monuments. Such was an unprecedented enterprise which was in part responsible for the pressure to begin farming. Megalithic architecture must thus be added to the question. Why start building huge stone structures? Obviously, there are folks who believe that sound and music may have had something to do with it. Something very interesting happens when we add a soundtrack to the picture, which is exactly what this book aims to do.

"The Sapient Paradox: can cognitive neuroscience solve it?" This was the way the question was posed by Professor (Lord) Colin Renfrew when the McDonald Institute for Archaeological Research at the University of Cambridge initiated a conference devoted to the theme "Archaeology meets neuroscience".[1] As Prof. Renfrew put it, "There seems to have been a long—in fact, inordinately long—delay between the emergence of anatomically modern humans and our later cultural flowering." Such a meeting as the one at Cambridge would certainly involve the *Megaliths* and *the Mind* part of our title.

A **megalith** is commonly described as a large stone that has been used to construct a prehistoric structure or monument, either alone or together with other stones. Stonehenge is the most iconic, although its sarsens are among the youngest of such compositions.

The mind part would be neuroscience, defined in *Psychology Today* as: *Areas of the brain. How brain cells work. Fields of study in neuroscience. The brain and common psychiatric disorders. Mental health treatment and the brain.*

Archaeology and neuroscience are an admirable combination which would definitely enlighten us about the mental acuity of people in the Late Stone Age. In the face of what the reader is about to learn, however, the scope feels incomplete, like two legs on a tripod.

What about the ***music***? Adding Archaeoacoustics to this mix introduces the transdisciplinary study of the human experience of special sound in ancient ritual and ceremonial spaces.

Researchers are now filling in the gap of knowledge about the psycho-physiological impact of music and certain resonant sound in the brain. There seems to be a trigger of some sort in sound with characteristics which were present in the world's oldest monuments. Added to new discoveries in Anatolia a solution for the "Sapient Paradox" practically leaps right out of the stone.

Sound is a subject that cannot be photographed or handled. Research requires input from a range of disciplines combined with informed empirical observation, weaving together threads that range from archaeology, architecture, cultural anthropology, ethnomusicology, genetics, physics, sociology and more. In composing the whole scenario, sometimes a little bombshell appears, destined for later elaboration. Instead of following a straight line with everything in its ordered place, the logic of this examination is more like a braid that winds back on itself. Welcome to the prehistoric spiral pattern of thinking. (Our brains can handle it.) If we adjust to it and include sound, it's a bit like adding missing colors to a painter's palette. Suddenly, the mix makes everything on the canvas richer and more vibrant.

6 | Introduction

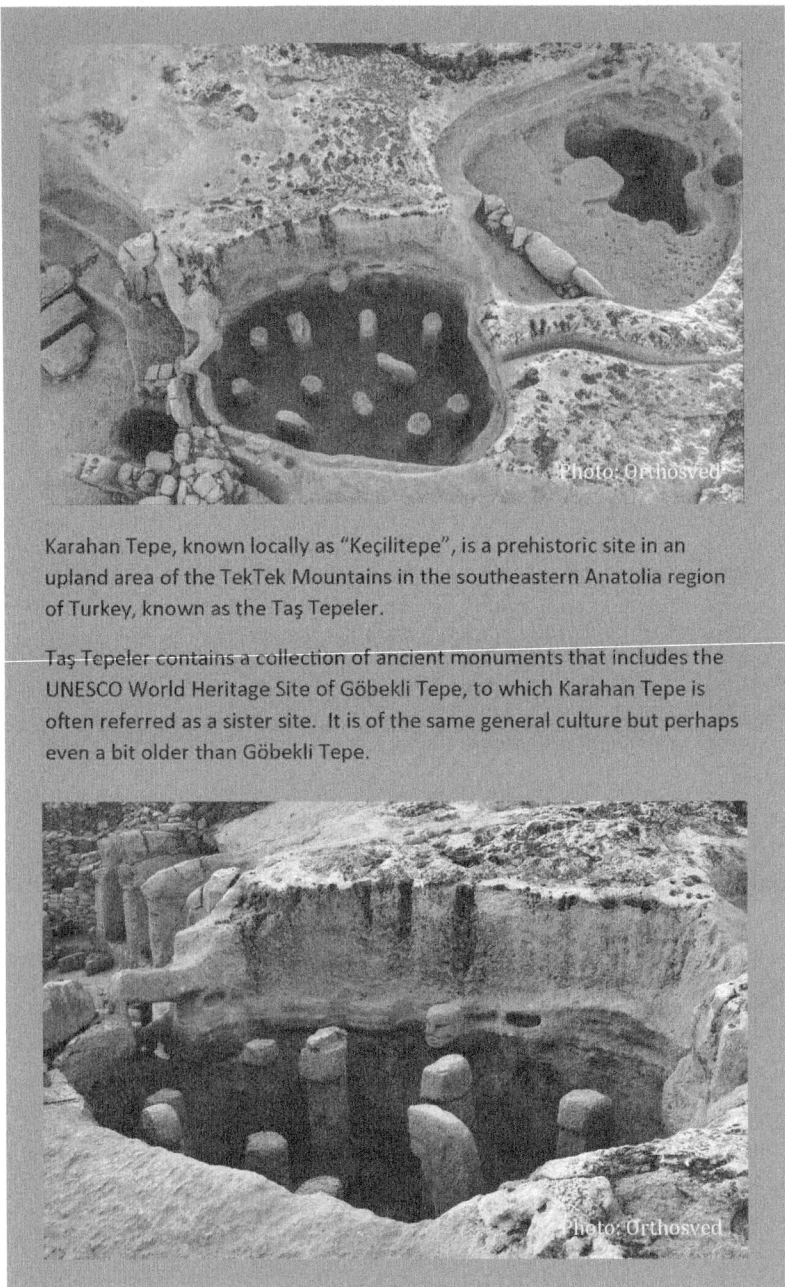

Karahan Tepe, known locally as "Keçilitepe", is a prehistoric site in an upland area of the TekTek Mountains in the southeastern Anatolia region of Turkey, known as the Taş Tepeler.

Taş Tepeler contains a collection of ancient monuments that includes the UNESCO World Heritage Site of Göbekli Tepe, to which Karahan Tepe is often referred as a sister site. It is of the same general culture but perhaps even a bit older than Göbekli Tepe.

Figure 1.2 Karahantepe, Turkey, ca. 11,500 years old.

As with most inspired invention, there was one key moment in time. Let's try for a moment to imagine how it may have unfolded. The following speculative scenario helps to encapsulate the "story" that now requires so many pieces to tell. It is presented with a nod to plausibility elaborated by the information contained in later chapters. Readers are invited to make their own speculations when this book is finished.

Suppose that around 12,000 years ago . . .

People of Asia Minor are hunting and gathering in family groups or small tribes. They know all about caves. Many of these people have been using limestone caves for conducting their special rituals. They have noticed that there are certain areas inside where the spirits of the stone speak back to them in a weird way: echoes that they don't have the science to explain even if they cared to. These people are living very close to nature. They have a belief system that is probably a lot like other indigenous people who live off the land like the Native American Indians, rainforest tribes and Australian Aboriginals. They hear these echoes as something that is coming out from the earth.

One tribe has a favorite cave that they visit regularly during their nomadic wandering cycle. Because this cave is limestone and has a gathering area shaped just the right way, the echoes are really exceptional. The singers in this group like to use it very much. In this imaginary tribe, an elder who is a holder of knowledge has always been singing special songs in caves; caves that echo and make his songs sound even more important. After all, there are songs for everything important in life. There are working songs and secret songs of the initiated. Maybe it is a Holy Man or Medicine Man sort of character who does this singing because his voice "wakes up the spirits" in the strongest way. In fact, the pitch of his voice might have been a factor in his qualifying for his position.

There are some issues with this favorite cave. Although the "magic" is strong, the space may be difficult to get into or occupied by too many unpleasant creatures. Somewhere along the line, his brain stimulated by all the singing in the cave, one such spiritual leader (or perhaps one of the listeners) starts to envision a wonderful possibility. Soon, while observing the children playing with stones, an idea takes solid form. Small rocks would be easier to stack, but they wouldn't make the right sound. It would take big stones . . . huge stones . . . mammoth stones!

He/she talks to the men of the tribe. What this person is imagining is going to take a ridiculous amount of very hard work. He/she has got to motivate these folks somehow; really motivate them in a big way, because it will probably take more than a hundred men to do it. They'll need to call in help from other tribes. But the spiritual leader has been convinced. The others come to believe it should be tried, so they try it and they succeed.

From the time the first stone rises upright into the air and breaks into the realm of humans, awe and pride and a tremendous sense of creative power instantly change the relationship of the builders with their environment. The sky is now filled in a space where just moments ago it was open. They have reshaped the earth and changed the space around them in a mind-boggling way, and they have done this thing with their sweat and their cleverness.

They are encouraged to keep going and the results are most spectacular. One by one, great stones are assembled to enclose a tidy space as it is desired, as opposed to rough uneven caverns as they are found. Soaring decorated stones cast shadows and echoes. Even before it is completed, they know. Properly invited, Spirit of Earth and Life will occupy this space. It has become something utterly fantastic.

A man in a painted cave becomes a man in a built cave with the creatures carved into stone instead of painted upon it.

It reads like a movie. Nevertheless, it might have been something like that. Someone had an idea and that was enough. As will be shown, the most astounding things were accomplished with no money, no metal and no wheels. Knowing what we know in our world today, our first instinct might be to say that such things are impossible. In the absence of anyone telling ancient builders that the job couldn't be done, however, it *was* done. Readily available resources were enhanced with the simplest of tools and rearranged with ingenuity, with teamwork and with confidence.

In the time of their use, visitors to the monuments that are currently being excavated in southeastern Turkey were inspired with an unprecedented concept. It must have been powerful and convincing because it was then copied hundreds of times. It was painstakingly preserved and it was remembered. The know-how was passed down through many generations.

Just imagine a path of human development that starts with chasing a sound buzz and leads to the advanced technology and globalization that we know in the 21st-century. *Something* happened. Here we are to show for it; walking talking evidence.

A sound buzz? Unexpected, isn't it.

Shouldn't it be an archaeologist writing a book like this? Perhaps, but none has done so. Archaeologists are highly trained to describe what they see. Many either teach or specialize in one area of study, one location, a single time period and in great depth. That becomes his/her professional signature area of expertise. This is the way our 21st-century systems work, with an ever expanding network of pigeonholes of specialization. Whether in a classroom or out in the field, few archaeologists have time to read a physics journal or explore genetic or neurological research, however much they might like to do so. The dimensions of our scientific knowledge are just too vast. Unless the archaeologist is focused on a trail of investigation that takes him/her into these other realms of science, it is daunting to learn the codes, abbreviations and special language that so many fields have developed.

We people of the modern world have had to compartmentalize our knowledge to such an extent that "the forest has disappeared because of the trees." Taking a different tack such as the multidisciplinary aspect of the study of Archaeoacoustics has the power to bring us back to a more personal relationship with the ancient past. Moreover, we have a chance to use it to our benefit in the modern world. As will be shown, the sentiment of "wisdom lost in knowledge" could hardly find a more palpable application. It is hoped that each reader will make some discovery inspired by a fresh perspective or a thought provoked.

The study required to cover this subject in the present must come from an institution which can umbrella multiple departments. It benefits from having a spearhead to oversee the various moving parts. To date, the OTS Foundation for Neolithic Studies has been that institution. In the unfolding of activities on behalf of OTSF for over 30 years, the author counts herself extremely fortunate to have had the opportunity to meet in the field with archaeologists, anthropologists, historians and other professionals, and to nudge the door open on this amazing subject. Presenters at the international OTSF conferences on Archaeoacoustics have brought the foundation of this book. The reader will meet several of them in this volume.

Those referenced conferences followed a strange chain of events.

1.2   Emergence of a New Field of Study

The story begins with a sequence of accidents and curiosities. When an English writer and artist (Paul Devereux) and a Princeton physicist (Robert Jahn) met in the ruins of a great round Anasazi kiva at Chaco Canyon in New Mexico, USA, they came up with a plan that would result in an evaluation of the acoustic properties of several ancient megalithic sites in England and Ireland. Modern techniques for measuring sound behavior that are part of acoustic engineering and architectural design were applied to Stone Age built spaces. The intriguing findings were shared in a workshop at a Cambridge University archaeological conference on methodology. An English television program emerged, and was eventually picked up for the in-flight entertainment program on a British Airways trans-Atlantic flight.

The author was a passenger on that airplane, and already had an interest in the subject of the Neolithic Stone Age as well as familiarity with an intact prehistoric "built" site that had a reputation for remarkable acoustic anomalies: the Hal Saflieni Hypogeum on Malta. Bored in a westbound full-daylight cabin, she scanned the available entertainment offerings, tuned into "Secrets of the Dead: Sounds from the Stone Age" produced for *Channel Four* and was immediately engrossed. No writing paper. She tore the cover off a paperback novel and started scribbling notes. Back in the office, they went into a file for "someday".

Sometime around 2002 or 2003 "someday" arrived. There was a concept in development with MacGillivray Freeman Films for production of an IMAX® Large Format Film about the Stone Age monuments of Malta and this research was a perfect fit. The author went to Princeton to talk with Bob Jahn and his associate Brenda Dunne. Unfortunately, that motion picture was not made. (There remains hope that one day the right mix will materialize for documentary production.)

The subjects of sound in antiquity and advisability of an acoustic evaluation in Hal Saflieni were anyway introduced where possible in public media. Exposure on the worldwide web churned up unpredicted interest. Response to internet and US newspaper articles demonstrated a hunger for more data and a suspicion that the study of Archaeoacoustics might present a chance to recover something akin to "ancient knowledge" that had been lost over time. The wheels were turning.

1.3 Assembling Expert Voices

Flash forward a decade. Scholars, researchers and enthusiasts from five continents collected in a conference room at the Corinthia Palace Hotel in Malta, where each of them had a provocative eye- and ear-opening experience.

As cited in 2004 by Professor (Lord) Colin Renfrew, Senior Fellow of the McDonald Institute for Archaeological Research: "For the prehistorian, Malta is one of the most remarkable places on earth . . . Something really exceptional was taking place in Malta more than 5,000 years ago, something quite unlike anything else in the Mediterranean world or indeed beyond."

Something exceptional again took place in Malta in February of 2014.

Such a gathering would have been unthinkable without the modern magic of the internet. In the spirit of Malta's prehistoric temple builders, a community of dedicated souls from many fields of study came together and created something both monumental and unforgettable. Their goal was to focus in a responsible way on the behavior of sound in important ancient spaces, and the way that ancient people may have used it. They sought hints for the way sound may have impacted on early human development. They applied a broad base of expertise, science, and objective observation toward a multifaceted understanding of human ingenuity. Evidence for the ancient use of sound in monuments and sacred places was far more widespread than previously believed.

Fresh insights came from putting all these people in the same room. The archaeologist had something to learn from the architect; the musicologist could inspire the anthropologist; the physician might collaborate with the sound healer; the art historian could get explanations from the engineer. Participants learned from a psychologist about the human experience of a primordial scream. They found that Babylonians sang out painful emotions in a language as fierce and angry as any modern rap lyric. And they had a chance to sample sounds in an environment that echoes and astounds today just as it did more than 5,000 years ago.

As a result of that inaugural conference, the subject of Archaeoacoustics was elevated from the perception of "pseudoscience" to a legitimate emerging field of study. Contacts were made for future multidisciplinary research collaborations. Specialists in one field were respectfully given a look into the mechanics of another, toward a broader understanding of the ancient world. New friends were made. Long years of work were validated. A roadmap had been drawn for future study of the archaeology of sound. A second conference unfolded the next year in Istanbul. And in due course, by popular demand, a third was conducted in Tomar, Portugal.

The pages that follow include input distilled from published papers, private conversations and a vast amount of far-flung research. In-house notes have been expanded, illustrated and augmented through the kindness of various experts, to whom the author is indebted.

In the conducting of these three conferences, it has become clear that ancient people all over the world have incorporated sound in ways that go far beyond the singing and music making that were part of every culture we know. It is not intended in this book to undervalue the Aztec sound heritage, or singing columns in ancient temples of India, or even the practice of sound in the Eleusinian Telesterion of ancient Greece by not including them here. There are loads of papers in the conference proceedings. Rather, the present goal is to step all the way back to the beginnings.

For reasons which will become clear, this book focuses on the megalithic traditions of the Mediterranean and Levant that were launched around 12,000 years ago, after the end of the last Ice Age. Concerns with a built acoustic environment go hand-in-hand with the so-called Neolithic Revolution which manifested for the first time in that area. This book does not try to separate them. We'll call it "lateral thinking." As will be explained, we might all still be living as nomad hunter/gatherers if the building of great stone monuments had not changed things.

The human minds of that time processed data in exactly the same way that we do today, but their interpretation of that data was vastly different. Lifestyle, social structure and core values were all very different. Most modern people cannot even imagine it without very conscious effort. We must accept that we have thousands of years of religious and political filters installed that color our perceptions. The challenge to the reader who wants to understand prehistory is to try to suspend them.

We're going to use the oldest man-made spaces on earth to explore the archaeology of sound. If we seem to spend an inordinate amount of time on Malta, the reader will soon understand why. He/she may even confirm something that was always suspected somewhere deep in the subconscious. In the words of world-class opera performer Joseph Calleja: "We might be in for a surprise to see how our ancestors dealt with sound, with amplification, with sound manipulation."

The study of ancient sites which are still enclosed is fundamental for examining Archaeoacoustics. Augmenting the 2014 work in Malta, the British Isles seem to be in the lead for collecting resources and applying acoustic research in the field. The internet provides many references these days. One example is acoustic experiments carried out at the chambered cairns of Camster (Caithness, Scotland), which revealed some very interesting results. Researcher Aaron Watson reports, "Sounds generated inside Camster Round appear loud and enhanced because they are contained by the confined stone walls. In contrast, sound does not travel easily along the passage, so a listener outside will hear a distorted impression. Higher frequencies are lost, emphasizing sounds such as drumming.

"Moving away from the passage to the sides of the cairn, higher frequency sounds become even more filtered. Around the back of the cairn, bass sounds created by drumming in the chamber sounded like they were emerging from beneath the ground rather than from inside the cairn.

"Intriguingly, the sound of a drum being played within the chamber of Camster Round could be distantly heard from within the chambers of nearby Camster Long. These monuments are 200 metres apart, and the sound could not be detected in the open air between them. One explanation is that these sounds are being transmitted through the ground. It is also possible that the silence in the chambers makes it possible to hear sounds that were otherwise obscured by natural noises in the outside world."[2]

Doubtless, as archaeoacoustic research continues, we will hear more about these sonic tricks from the past that have been hidden in ancient monuments.

As regards the listing of dates, in the interest of minimizing confusion stemming from a system that often requires cumbersome adding and subtracting from/to AD, BC, BCE, YBP, every attempt has been made herein to give ages using a uniform *years ago*. There remain a few places where traditional dates have been kept for citation reasons and in quotations.

Figure 1.3 Mnajdra Temple Complex, Malta. Filming an ancient (late 1990s) interview with the author for *Discovery International*.

Moving onward, we come to an independent island republic about 60 miles south of Sicily. Only now emerging from under the radar on its own merit, the islands of the Maltese archipelago (principally Malta and its sister island Gozo) have appeared on the big screen for many years. A combination of factors make this location ideal for filming. The long list of major motion pictures shot in Malta includes "Popeye", "Midnight Express", "Gladiator" and "Troy", as well as the first season of "Game of Thrones". In addition to the fantastic light that is reflected from an abundance of pale ivory limestone and so much surrounding sea, one feature that lures filmmakers is the variety of Malta's architectural legacy in stone. Over a long history, the islands have been visited or occupied by every major force that ever sailed Homer's "wine dark" Mediterranean. Each of them used the same beautiful stone to build in their own particular style. Consequently, there are places in Malta that can stand in for just about anywhere and fit the desired historic period, be it Roman or Medieval, WW2 (or mythical Atlantis as it did in "Eric the Viking"). Our interests, however, lie in the prehistoric.

On this rocky little island, we chasers of ancient sound hoped to move beyond theory and the one-dimensional artificiality of electronic signals. We wanted to try to sample the real deal. A select team of participants was assembled to partake in a site evaluation: a pre-planned opportunity that had been years in the making. It did not disappoint.

We are not desolate—we pallid stones;
Not all our power is gone;
Not all our fame;
Not all the magic of our high renown;
Not all the wonder that encircles us;
Not all the mysteries that in us lie—

Edgar Allan Poe, 1835

Chapter 2

The Ħal Saflieni Project

The world's oldest freestanding buildings: the twin Ġgantija Temples and other megalithic structures like them on the islands of Malta and Gozo were already more than a thousand years old when the great pyramids were built. Imagine them—already ancient and mysteriously abandoned when the stone circle at Stonehenge was begun. To find anything older, one would have to look about 1,350 miles (ca. 2,200 km) to the east, to excavations like Göbekli Tepe in Turkey. Until recently, those fantastic post Ice-Age stone shrines about which more will be discussed, had remained hidden in the earth where they were intentionally buried for more than 9,000 years. When Malta's temples were built, they were the only visible standing monuments in the world.

Although largely under-reported in the popular mainstream, the cultural remains of Malta's "Temple Builders" comprise the richest and arguably the most intact collection of Neolithic society surviving in the world. Magnificent artifacts, pottery, undisturbed burials, and breathtaking megalithic monuments with walls still enclosing space around original pavements are an enthusiast's dream.

Megaliths, Music, and the Mind
Linda C. Eneix
Copyright © 2024 Jenny Stanford Publishing Pte. Ltd.
ISBN 978-981-5129-25-0 (Hardcover), 978-1-003-46823-3 (eBook)
www.jennystanford.com

Figure 2.1 Megalithic Temples of Gozo and Malta, from ca. 5,800 years old.

There are more than 23 places on the islands where it is known that megalithic temples once stood. Four of them are in exceptional states of survival.

Everything one can read about the society that built them will say that the settlers came from Sicily. Although the implication that they were Italians is ridiculous, the original Neolithic people probably did sail from there with their livestock, seeds and plants. Sediment cores drilled in Malta show at that time period the introduction of pollen from nettles and olives in addition to the expected grains. Domesticated olives, which are larger and juicier than wild varieties, are said to have been first cultivated six to 8,000 years ago from wild olive trees at the frontier between Turkey and Syria where Göbekli Tepe lies. Nettles produce fiber that becomes an excellent textile if one has the patience to process the stems. Artifacts of the period do show evidence of textile weaving.

The prehistoric colonists didn't choose Malta because the farming was good. There has never been more than a thin covering of soil over the natural limestone of which the land was formed many millions of years ago when the force of the African tectonic plate pushed the islands upward on its way under the European plate. Even today in Malta and Gozo, dairy cattle spend all their lives in confinement in the islands' few dairies because there are no fields for grazing. Sicily is a far better place for farming things to eat.

The first people may have isolated themselves on the islands, at least in part, because of the wonderful stone. For more than a thousand years, they embellished and expanded the megalithic monuments that they raised here. Then around 4,500 years ago, the entire population disappeared rather abruptly and mysteriously. It is not known what happened to them; there are no signs of warfare or mass disease. In the archaeological record, there is simply no sign of human occupation for several hundred years. There is evidence in Malta's sediment cores of some enormous marine catastrophe, but this has not yet been elaborated. The monumental sites fell into ruin and started filling with ash, dust and debris until a Bronze Age society arrived and reused at least one of them for their own funeral practices. These newcomers left remnants of a completely different material culture.

It is rare to find such relics that can still suggest their former glory to us to such an extent. We might be able to imagine the structures very close to the way the original builders knew them. Yet, as could be expected after so much time and exposure, much of the original construction is missing from Malta's megalithic monuments. Ceilings are long gone, so we'll never hear sound in those shrines as it sounded to the people who built and used the places in prehistory. However, . . .

Also on the island of Malta, against all odds, there is a treasure of a place that was made by the same temple-building Neolithic society; one which is intact and out of sight. Basic human nature hasn't changed much in six millennia, and neither has Malta's Hal Saflieni Hypogeum. It is possible to drive past it and over it every day without notice. Under the streets and houses of a late 19th-century neighborhood lies a network of halls and rooms, sculpted from the living rock with architectural features that mirror the megalithic temples like Ġgantija above ground.

Katya Stroud, Malta's Senior Curator of Prehistoric Sites, underscores the significance of the site with regard to the development of human processes of cognition and representation. ". . . The rock cut chambers of Hal Saflieni allow us to study a system of interconnected spaces very much as they were conceived by a Neolithic mind."

The word hypogeum comes from ancient Greek terms for a building under the ground. Technically, it can be applied to any vault or subterranean structure, but if you say it in Malta everyone thinks of only one place. *Hal Saflieni* and *The Hypogeum* are used interchangeably when discussing this site.

During World War II, the British kept much of the Mediterranean fleet in the naturally deep-water creeks of Malta's Grand Harbour. It is reported that 15,000 tons of bombs dropped on the Maltese islands, and most of them were in the area that housed ships and ammunition: the part of the harbor that is just a few blocks walk from Ħal Saflieni. Some might say providentially, the site somehow remained both architecturally and acoustically intact.

Like the Göbekli Tepe remains in Turkey, this remarkable place had not been disturbed during all the thousands of years from its disuse until its discovery. Ħal Saflieni remained pitch dark and silent through every event in history—from around the time the pyramids were being built until 1902, when it was accidentally broken into during construction works in the neighborhood above it. Legends and myths about its sound behavior have persisted since that discovery.

In all that time, Ħal Saflieni never collapsed or fell into ruin. There had been no re-making of the space by Phoenicians or Romans or any other culture that occupied Malta over its long history. Nobody came along and dismantled it to reuse the stone. No one smashed it because it was a symbol of a defeated heresy. Nobody even knew it was there.

As an architectural structure, it is very much as whole now as it was when it was abruptly abandoned, along with its companion freestanding megalithic temples, at just about the time that Egypt was getting its great pyramids at Giza. The soundscape in Ħal Saflieni is virtually unchanged from what it was nearly 6,000 years ago. Sound behaves within this space exactly the way it was heard by people in the Stone Age. The site has been a springboard for the launch of our exciting new study.

Ħal Saflieni is an extraordinary place inscribed on the UNESCO World Heritage List as "a site that bears a unique testimony to a cultural tradition which has disappeared."

It was created as an underground necropolis, and is a tribute to the artistic, engineering and architectural skills of Malta's prehistoric "Temple Culture". Although there are known to be more subterranean complexes like it on Malta and Gozo, Ħal Saflieni is the principal one, and rightly claimed by Heritage Malta to be the only prehistoric burial site which is accessible to the general public.[3]

The composition of the space confirms that in the Neolithic (New Stone Age) period, Ħal Saflieni was used not only as a depository for bones, but also as a shrine for ritual. Whether that was restricted to use exclusively by an initiated few or periodically for mourning family groups or regularly for as much of the community as could be fit inside, we will never know.

FAIA architect Prof. Richard England was once asked how he would go about rebuilding such a place today. He responded: "I don't think we could. We might have the technology but we don't have the soul..."

The complex is made up of interconnecting rock cut chambers set on three distinct underground levels. The earliest remains at the site date back to nearly 6,000 years ago. It was in continuous use over a span of about 15 centuries. (The united Roman Empire lasted less than five.)

The uppermost level, consisting of a large hollow with smaller chambers on its sides, is now covered by a climate-controlled visitor center. According to archaeologists, this hollow was probably originally exposed to the sky. Excavations in the early 1990s indicate that there might also have been a monumental structure of some kind marking the entrance.

A portal leads down to the middle level, which contains some of the best known features of the Hypogeum such as the beautifully carved architecture in the negative. Several of the ceilings in this area are exceptional. The sculptural modeling of the space has reproduced a fine example of the corbelling

Figure 2.2 The above three images are of the Ħal Saflieni Hypogeum, Malta, ca. 5,800 years old.

Figure 2.3 Ħal Saflieni Hypogeum, Malta, ca. 5,800 years old.

Figure 2.4 Ħal Saflieni Hypogeum, Malta, ca. 5,800 years old.

that is still found in the remaining ruins of the temples, as well as a feature that residential architects call a tray ceiling. This is our guideline for guessing how the temples were once roofed and how their interiors must have looked originally. Intricate paintings in red ochre decorate some of the walls and ceilings. A lower level, accessed down seven treacherous stone steps, is a subject for conjecture, but not our primary concern here.

The author's first visit came in 1990 when one needed simply to make a phone call, pick up a key and borrow a flashlight. Imagine the thrill of being able to wander at will in such a place: timeless and surreal. Today, there is a computer-controlled conservation system that completely changes the air every 60 minutes. Ten people per hour are permitted to visit below. For those lucky enough to get a ticket, the first indication of the sound environment of Hal Saflieni is the echoing footfalls that follow the site curatorial representative on the metal entry stairs that lead to a railed visitor walkway. Very soon thereafter one is engaged by a pre-recorded narration to guide the tour. To control algae growth, the lighting is dim and timed to come on and off again as the visitor is ushered through the space with his talking recorded guide. Tourists are usually highly impressed by the site itself. They frequently exit speaking to each other in whispers if at all. Most visitors leave without being aware of the significance of the sound environment.

For the Archaeoacoustics researcher, visiting Hal Saflieni is the next best thing to time travel. It is a complicated site that is not easily defined acoustically. Many have tried it. Some of the research will take a long time to analyze, although preliminary findings are indicated later in this book. Certainly there is room for much more sophisticated acoustic evaluation work.

Suffice it to say that the sound quality in Hal Saflieni can raise the hairs off the back of the listener's neck. A low voice or drumbeat wraps around the space and doubles back on itself. One cannot tell exactly where the sound is coming from and yet it is felt in the deepest tissues. The experience can be imagined as standing inside a giant cast iron bell as it's struck. One's ears are tickled with feedback. The air is vibrating; the walls and floors and ceilings are vibrating; the sound is swimming all around and one can feel it in the tissues of the body. There can be a sort of buzzing in the ears that sometimes comes when singing along with the radio in the car—when one has exactly matched a pitch. Sometimes there is a residual rhythmic "thrumming" after an echo has faded and silence returns, as when someone runs a wet finger around the rim of a crystal glass. It could be the listeners own heartbeat. It's mesmerizing. And oh yes, there is an addictive quality. *More. Do it some more.*

Workers in the space, even talking to each other, would have been exposed to the acoustic effects. It is difficult to imagine that human beings would let the potential of a sound phenomenon such as that in Ħal Saflieni go unexploited. If it's impressive today, for people who seem to have answers for everything, the thought of what they might have created in here thousands of years ago is extraordinary. Were they listening for the voices of the dead, or perhaps spirits of the earth? There would have been an aroma, as well as an echo, creating a very sensory and emotional experience. We now know that while the goosebumps were rising, something was happening in their brains.

More than 5,000 years ago visitors to this site shared it with the bones of their ancestors which had been ritually placed inside the earth, although the place was far more than an ossuary or tomb. Quite a bit of the architecture in Ħal Saflieni suggests features of our performance spaces today. Considered with artifacts recovered from this site, it is clear that something more than interment of the dead was going on in here.

It is in the middle level that we find the chamber popularly known as the "Holy of Holies", as well as a compartment section that has been called "the Oracle Room". This is the chamber artfully described in 1920 by William Arthur Griffiths in *National Geographic Magazine*. "A word spoken in this room is magnified a hundredfold and is audible throughout the entire structure. The effect upon the credulous can be imagined when the Oracle spoke and the words came thundering forth through the dark and mysterious place with terrifying impressiveness."[4] (For readers who would care to hear it, an audio sample is tucked below a contact form at https://otsf.org/targetinginitiative. Use headphones or a good speaker system with a subwoofer if possible.)

This eggplant-shaped "Oracle Room" functions as a giant resonator. There is a high carved out indentation curving around the dead end of the chamber that suggests no purpose but to channel sound waves. The remains of red ochre on the ceiling in this chamber are an intricate pattern of disks and curls that begin above a side compartment and spin out like some kind of prehistoric musical notation, stopping at the entrance portal that frames more finely carved halls beyond.

Several side compartments were cut into the side wall of the chamber. One of particular interest is at face height. There are three red discs painted on the inside, which we archaeoacousticians think may have been targets for aiming vocals. One can see discoloration on the lower rim of the niche caused by countless hands that have rested there while their owner leaned forward to place his/her face inside (just visible above the elbow level on the left in the illustration; see Fig. 2.6). It has been said that a male voice speaking into this compartment is heard throughout the three stories of the underground complex in a way that is far different from a female voice. The phenomenon has to do with physics. Like Newgrange Passage Tomb and every other tested megalithic site, the range for standing wave resonance in Hal Saflieni responds to a very deep tone. More on this in a moment.

The Ħal Saflieni Project | 33

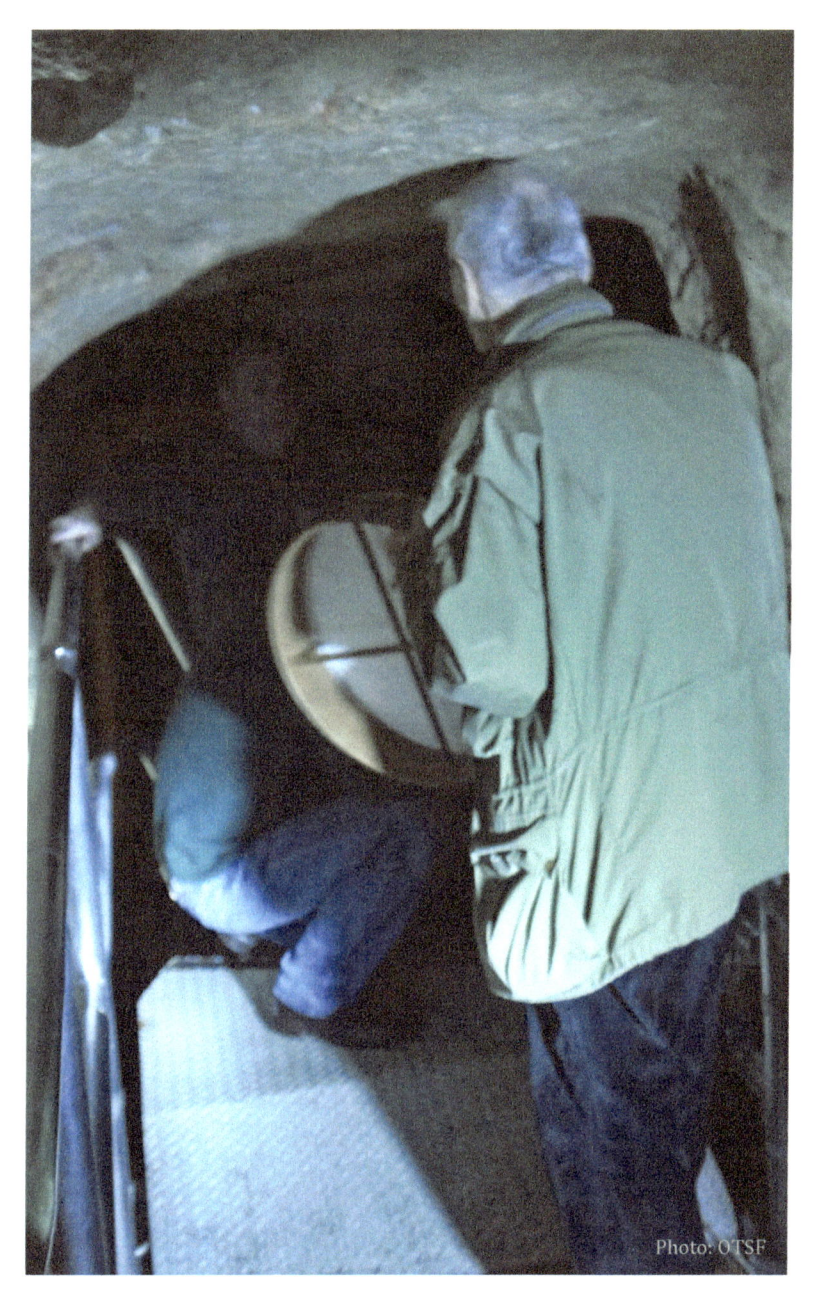

Figure 2.5 Music Anthropologist Prof. Iegor Reznikoff and Rock Art Specialist Fernando Coimbra enter the "Oracle Room" prepared for archaeoacoustic evaluation of the site.

Figure 2.6 Ħal Saflieni's "Oracle Room" with it traces of red ochre design on the ceiling.

Voice and sound have been part of nearly every religious ritual across time and cultures. This is because music has the power to touch our souls. It is safe to say that it was the same for our distant ancestors, and for the original users of Hal Saflieni, maybe even more so. They had all the emotions that we have. They had the same vocal cords and the same rhythms. They had the exact same natural propensity for making music and drama that we have.

While it is important to note that we can never know for certain what transpired within the Hypogeum 5,000 years ago, science can help expand the possibilities. The February 2014 exercise is the first scientific examination of its resonant properties to produce publishable results.

Our investigative team was richly multidisciplinary. Archaeology, Architecture and Acoustic Engineering helped define the *when, what* and the mechanics of *how*. Backgrounds in Art and Music helped examine abstract expression. Psychology and Anthropology addressed a human element that has not changed in long millennia.

2.1   The Megalithic Range

Under the supervision of Heritage Malta, we placed microphones in the Oracle Room chamber. Two digital recorders were used (Tascam DR-680 and Tascam DR-100) with power battery set at 192 kHz of sampling rate for DR-680 (range 10 Hz to 96 kHz) Ultrasensitive microphones: Sennheiser MKH 3020 (frequency response of 10 Hz to 50,000 Hz) with shielded cables (Mogami Gold Edition XLR) and gold-plated connectors. The response of the chamber was tested during stimulation by different voices and by simple musical instruments which could have been present in the time the Hypogeum was in use. It was possible to stimulate the maximum phenomenon both with a low male voice and with several drums tuned to very deep notes.

We all know that sound is vibration, but beyond that the particulars can be murky. The following very simple explanation comes from *Science World*.[5]

> "Three things vibrate when sound is created: the source object, the molecules in the air (or another medium, e.g., water), the eardrum.
>
> "When a sound is produced, it causes the air molecules to bump into their neighbouring molecules, who then bump into their neighbours, and so on. There is a progression of collisions that pass through the air as a sound wave.
>
> "Air itself does not travel with the wave (there is no gush or puff of air that accompanies each sound); each air molecule moves away from a rest point and then, eventually, returns to it.
>
> "When we hear something, we are sensing the vibrations in the air. The number of vibrations per second is known as the frequency, measured in hertz (1 Hz = 1 vibration per second).
>
> "These vibrations enter the outer ear and cause the eardrum to vibrate too. We cannot hear the vibrations that are made by waving our hands in the air because they are too slow. The slowest vibration our human ears can hear is 20 times a second. That would be a very low sound."

Most healthy adults have an average hearing range of between that 20 Hz on the low end and 20,000 Hz for sound that is high in tone. For comparison, a dog hears somewhere between 40 to 60,000 Hz. The range for a cow is 23 to 35,000 Hz. A dolphin hears sound from an amazing 2 to 200,000 Hz.

In Hal Saflieni, our team of investigators detected the presence of a strong double resonance frequency. They measured spikes at 70 Hz and again at 114 Hz. The numbers varied slightly with another mechanical method of testing which confirmed the presence of multiple "peaks" in a range between 41 to 140 Hz. By a third method, an expert using his well-trained voice cited 82 and 110 Hz.

In the overall scheme of things for the purposes of this study, a few hertz is not a critical difference. Prehistoric folks would have found the numbers useless anyway. Essentially, in the presence of sound within the register of a low baritone or bass singing voice, all sorts of weird things start happening in the Hal Saflieni Hypogeum acoustically. More will be said about this in the section on Music.

Actually the acoustics of Hal Saflieni are even more complex, with loads of low frequency vibration that is beyond human hearing, although one might be able to feel it.

In physics, resonance is the tendency of a system to oscillate at a greater amplitude at some frequencies than at others. These are known as the system's resonance frequencies. At these frequencies, even small periodic driving forces can produce large amplitude oscillations, because the environment's acoustic system stores vibrational energy. To quote Britannica.com: "An example of acoustical resonance is the vibration induced in a violin or piano string of a given pitch when a musical note of the same pitch is sung or played nearby."[6]

Essentially, this is what happens in Hal Saflieni. When stimulated by voice or drum at the resonant frequency of the site, the place itself responds with vibration. The hard surfaces reflect that vibration and the echoes chase themselves around the curves.

Glenn Kreisberg is a radio frequency spectrum engineer who was with the group. His perception of the space from his area of experience is worth noting further.

"Upon entering the Hypogeum I was struck by the curved nature of the walls, pillars, stairways and ceilings, with no sharp corners, edges or surfaces. Not unlike the above ground temples of Malta, many of the lines are circular, which creates numerous, continuous, opposing parallel surfaces. Everything seemed carved or worn smooth, almost as if slightly polished, perhaps the result of water running over the surfaces for so a long period of time. The effect on propagating (sound) waves this curving smoothness plays is that it prevents refraction or the bending of waves when they encounter a sharp or jutting surface. Think how ocean waves "turn" or bend when encountering a jutting peninsular, as they brake toward the shore. Refraction breaks down and weakens waves (water, sound or EM) and is unlike reflection. The many parallel, opposing surfaces of the Hypogeum cause reflection, which allows the sound waves generated within to echo, build upon themselves and reverberate strongly."[7]

While we found that natural horns and vocals in a higher range certainly made noise, the low male voice and drum tuned to these frequencies stimulated bone-chilling effects. It was reported that some sounds echoed for up to 13 seconds during testing in Hal Saflieni. Try to imagine how impressively long that is for a sound to persist before decaying to silence.

> **Exercise**
>
> Try to get a group (whose members can overlap each other to keep the sound constant while breathing as in chanting) to hum a note continuously for 13 seconds, gradually lowering the volume. A second group starting with another note 5 or 6 seconds after the first could be interesting.

Whether or not our prehistoric ancestors consciously manipulated space to stimulate this eerie sound behavior is something yet to be determined. The evidence will be presented and the reader can decide. In any case, they certainly were exposed to it. As we'll learn, if anyone was even simply talking in there with a low voice, exposure to the sound affected them physically. There is much that can be surmised even if we need to speculate the exact use they were making of it for their ritual worship or burial ceremonies.

It is naïve to think that the original people who used the place didn't notice something like the impressive sound effects that happen inside it. The natural response even of children in an echoing room is to make more echoes and play with the sounds they can make. Would the ancient users of Hal Saflieni demand absolute silence in the space? Or would they take advantage of the acoustics to make their ritual vocalizing or drumming sound particularly powerful and authoritative? Hollywood has been onto this concept for a long time. The voice of God is almost always delivered with an echo because that differentiates the important message of divinity from the mundane speech of mere mortals. It is also a universal practice of filmmakers these days to use the soundtrack to define and direct the emotional content of a scene or even an entire motion picture. Indeed, they have perfected this as an art form.

The resonance frequency numbers from the Hypogeum fit into the findings of tests made at other megalithic sites from Turkey to Ireland, all generally showing their strongest resonance in a narrow range of pitch. The first reports that involved archaeological sites came from the British Isles, where there are plenty of similarities in megalithic constructions.

Research in the 1990s by Archaeoacoustics pioneers Paul Devereux and Princeton Physicist Dr. Robert Jahn concentrated on the acoustic resonance of six prehistoric chambered "tombs" in England and Ireland. "The question was posed as to whether ancient ritual structures could have arrived at their proportions as the optimal result of empirical recognition of the acoustical properties of the kinds of ceremonial singing or musical sounds for which they may have provided the environment."[8]

All the tested sites, including Newgrange Passage Tomb, Cairn L, Loughcrew, Co, Meath and Wayland's Smithy, Berkshire yielded resonance frequencies within a narrow band between 90 and 115 Hz. These findings essentially herald the results of testing in the Hal Saflieni Hypogeum, although the latter site is far more complex.

It appears that these particular frequencies of the Hypogeum and stone cavities like Newgrange Passage Tomb and other cairns, have a strong physical effect on human brain activity.

2.2 Human Response

Dr. Ian Cook of UCLA and colleagues published findings in 2008 of an experiment in which regional brain activity in a number of healthy volunteers was monitored by EEG through exposure to different resonance frequencies. Their findings indicated that on exposure at 110 Hz, in the midst of this "megalithic range", the patterns of activity over the prefrontal cortex abruptly shifted, resulting in a relative deactivation of the language center and a temporary shifting from left to right-sided dominance related to mood and emotional processing.[9]

A strange modern mythology has emerged around this very preliminary sampling at UCLA. Sound healers from the USA to Europe to Australia have picked up 110 Hz as a magic number that relates to the measurement of the pyramids and all the workings of the universe. That might be so, but for the reader who is serious about delving into archaeoacoustic research, we stress the importance of reading the entire original scientific reports and not just pouncing on something found on the internet and taking it out of context, because the picture isn't that simple.

It is suspected that everyone has his own private "trigger pitch" within this range of about 70 to 120 or 130 Hz, at which the brain responds. Dr. Paolo Debertolis reports on tests conducted at the Clinical Neurophysiology Unit at the University of Trieste: "Each volunteer has their own individual frequency of activation, always between 90 and 120 Hz. Those volunteers with a frontal lobe prevalence during the testing received ideas and thoughts similar to what happens during meditation, while those with an occipital lobe prevalence visualized images."[10] While remaining in the cited range, there is an indication that the exact trigger point can vary slightly by age and possibly even in the same individual from one day to another. Debertolis goes on to summarize that under the right circumstances, "Ancient populations were able to obtain different states of consciousness without the use of drugs or other chemical substances."

An unpublished experiment at the University of Malta yielded similar results, as did a pilot test in the UK. Something is happening in the brain when stimulated by sound at around that level. Clearly, much more work needs to be done. A full study will take a dedicated lab, a wide range of test subjects and about a year's time. Later pages in this book introduce even more compelling brain work related to sound and music. At the very least, we could identify an emotional element. One can only imagine the impact in antiquity, of sitting or standing in near total darkness, listening to weirdly echoing ritual chanting while low light flickered over the bones of one's ancestors.

What is particularly astounding is that it seems the Hal Saflieni builders not only recognized that the environment was inclined to produce sound effects but they exploited it, intentionally using architectural techniques to boost the *super-acoustics*. Kreisberg observed that in the Hal Saflieni Hypogeum,

> "The Oracle Chamber ceiling, especially near its entrance from the outer area, and the elongated inner chamber itself, appears to be intentionally carved into the form of a wave guide.
>
> "Preliminary modeling based on photographs and floor plan diagrams of the Hypogeum, confirmed previous research which suggested that the room known as 'The Oracle Chamber' has characteristics that apparently project sound energy in a highly focused manner, allowing the audio waves to easily propagate to the other areas and rooms in the large, multileveled complex. Was there something specific, a feature or aspects of the design that could account for such an effect? Was it intentional and planned for in the original construction design?
>
> "A wave guide is a structure that guides waves, such as sound waves or electromagnetic waves," says Kreisberg. "There are different types of waveguides for different type of waves. As a rule of thumb, a waveguide needs to be of the same order of magnitude as the wavelength of the guided wave. So, high frequency, small waves require a small wave guide and low frequencies with larger wavelengths would require a larger wave guide. The very low frequency sounds that echo strongest in Hypogeum have very long wavelengths, thus the wave guide employed would need to be quite large. I believe The Oracle Chambers size itself is of the magnitude as to create the wave guiding effect upon the sound waves produced within the structure."[11]

The carving of the two niches which appear to concentrate the effect of sound, the curved shape of the Oracle Chamber with its shallow "shelf" cut high across the back, the corbelled ceilings and concave walls that are evident in the finer rooms of the complex are all precursors of today's acoustically engineered performance environments. If we can accept that these developments were not by chance, then it is clear that Hal Saflieni's builders also knew how to manipulate a desired human psychological and physiological experience.

One might ask: Why would they do that? The first answer is, simply: because it feels good.

In their paper "Fear and Amazement" from the 2014 Hal Saflieni Archaeoacoustics exercise, Archaeologist and Psychologist, Dr. Torill Lindstrom and Anthropologist Dr. Ezra Zubrow remarked: "We regard it as almost inevitable that people in the Neolithic past in Malta discovered the acoustic effects of the Hypogeum, and experienced them as extraordinary, strange, perhaps even as weird and 'otherworldly.' Sounds can create emotional effects with neuropsychological concomitants . . . If the Hypogeum was used for rites-of-initiation, its acoustic properties may have been instrumental in creating emotional seesaw-effects with resultant compliance and conformity."[12] (See also Bonus Material in this volume.)

It was made clear at the conference that launched these investigations that across cultures, distance and timelines, special sound is associated with sacred spaces: from Paleolithic painted caves in France and Spain to singing stone temples in India; from Mesoamerican codexes to classical mysteries and sanctuaries in Greece to sacred valleys in Elamite Iran. (Tibetan, Asian and other traditions remain to be shared, perhaps at the next conference.) Stated simply, it is human nature to isolate these hyper-acoustic places from mundane daily life and to place high importance to them because abnormal sound behavior that is bigger than normal, implies a divine presence. In whatever form, most of us need to believe in something greater than ourselves that has the power to fix what is wrong, guide our decisions, keep us safe and fed. Many of us yearn to commune with the supernatural and control our fate. As we will see, there is a logic to this assumption that ancient ritual spaces were providing an environment for doing just that.

We also seem to be hard-wired to try to explain mysteries. Some of the myths and urban legends about Hal Saflieni that have come from visitors' interpretation of this extraordinary place hint at a psychological impact and underscore a propensity for imagining the fantastic. With the tricks of sound that happen naturally, it is an eerie experience, inspiring colorful stories. Heritage Malta's Senior Curator of Prehistoric Sites, Katya Stroud has heard them all.

"There is an account that in the 1940s a British embassy worker, Miss Lois Jessup, went on a tour of the Hypogeum and persuaded a guide to let her explore a 3 ft. square 'burial chamber' next to the floor of the lowest room in the lower level of the site. She claims that after squeezing through this chamber she came into a large room; where she was standing there was a large cliff with a steep drop and the floor of the cavern could not be seen. Across the cavern there was a small ledge with an opening in the wall. According to Ms. Jessup, a number of 'humanoid beings' that were covered in white hair and hunched over came out of this opening. They raised their palms in her direction and a large gust of wind filled the cavern, extinguishing the light of her candle. She then claimed that she felt something brush past her. She apparently survived the ordeal to tell the tale. There are no signs of such a chamber in the lower level of the hypogeum. Its extent has been found and recorded in a 3 D model using laser scanning. On the other hand, reflections from a candle in the water-logged chambers in the innermost part of the lower level are likely to have made the experience quiet disorienting for someone who was new to the site."[13]

There is another story that relates a group of school children and their teacher visited the Hypogeum on an outing and entered the same burial chamber, which then collapsed while they were inside. Search parties could not conduct a thorough search for the children or their teacher due to the cave-in. The parents of the children claimed that, for weeks, they could hear the voices of their young children coming from under the ground in several parts of the island.[14] However, Stroud has checked this out: "Research shows that there are no local newspaper reports or accounts from residents about missing children."

Plentiful on the web are conspiracy theories about elongated skulls that were discovered at the Hypogeum, put on exhibit at the museum of archaeology in Valletta until 1985 and then removed and placed into storage away from the public eye. "These so-called elongated skulls have been the subject of theories related to a race of serpent priests, an ancient genetic mutation between different races, and countless others," says Stroud. "Temi Zammit, who was the first scientist to study these skulls said 'About ten skulls, however, were recovered in a state that could be measured. They are all of the long-headed type (dolichocephalic) with an average cephalic index of 71.8'. Being a doctor, Zammit was probably familiar with the standard varieties of skull proportions found within modern society and was referring to the medical standards of; Brachcephalic (broad-headed), Mesocephalic (medium-headed) and Dolichocephalic (long-headed) which are incidentally the standard varieties of skull proportions that we find in today's populations." The snakehead theory carries no weight, although the "Serpent Priest Theory" is a popular one that has already been used in some media descriptions of Karahantepe, a recently discovered site that will be introduced in a later chapter. Once again, the reader is encouraged to seek credible scientific evidence for such an interpretation.

These fanciful ideas frequently make it to US television in a popular show broadcast on the *History Channel*. An actor uses scripted content to fill in the gaps and say the things the visiting experts will not. The author's personal experience with this practice includes the show in which he suggested that back in the Stone Age, the Hypogeum may have been a sonic beacon for incoming alien spacecraft.

No.

Let's be clear. There is zero credible evidence of alien intervention or any earlier advanced technological civilization that somehow instructed the survivors of a global catastrophe. People who made the Hypogeum and Newgrange and Göbekli Tepe were every bit as intelligent and creative as we are. It is insulting to them and their descendants to suggest that outside intelligence was needed. It would, however, be fascinating to know their myths and conjectures about what they heard in the highly designed stone environments that they created. There is every indication that a sophisticated school of architectural and audiologic knowledge was already in place a thousand years before the Egyptians started building pyramids.

If we take a long hard look at the Neolithic Revolution and the waves of migration of the time, it seems likely that the Hal Saflieni Hypogeum is a rare surviving remnant of a rich ethnic tradition. Could Europe's megalithic buildings, spread across thousands of miles and thousands of years, be part of a cultural memory that has to do with sound?

Mostly dating to a time that parallels the end of the Neolithic building in Malta, stone monuments and tombs appear in Sicily, Sardinia, Corsica and many points to the West which reflect architectural features of the great Maltese megalithic shrines. Some "passage tombs", like Newgrange in Ireland and Gavrinis in Brittany are contemporary with the Malta buildings. The thing about these passage constructions that suggests something important is that their chambers are too small to hold more than a few people. The forecourt would be the focal point for visitors to gather and hear sound that was generated from inside. Perhaps the benefits of being immersed in the vibration of the "sacred sound" was reserved for a privileged few.

Drs. Lindstrom and Zubrow hint at a hierarchal purpose for the manipulation of sound. "The Neolithic itself was characterized by cultures focused on new invention...enormous collective collaborations over extended periods of time. For these large-scale projects of agriculture and building, social cohesion and compliance was absolutely necessary." It seems entirely possible that those who knew the secrets of sound would have held some power. That could explain quite a bit.

From the very beginning, humankind has been listening. We had to be; life and death depended on it. Hungry predators, wildfire, twig snaps and footfalls, wild game birds, flowing water, friend or foe—we were depending on our hearing in a way that can hardly be imagined today. Whether they intended to be or not, ancient people were exposed in their limestone caves and monumental structures to sound phenomena that were very different from what they were used to: sound that could have acted in the mind like a drug. How would they explain such a thing when they had no science to give a reason for it?

Before we can delve more deeply into the perceptions and reactions of early man to sound made within the spaces he created, we need to take a look at the where and when of what happened after the last Ice Age ended and the building of man's first monuments began: the Neolithic Phenomenon. We all stand to gain from considering the "big picture of humanity" armed with perspective that includes prehistory.

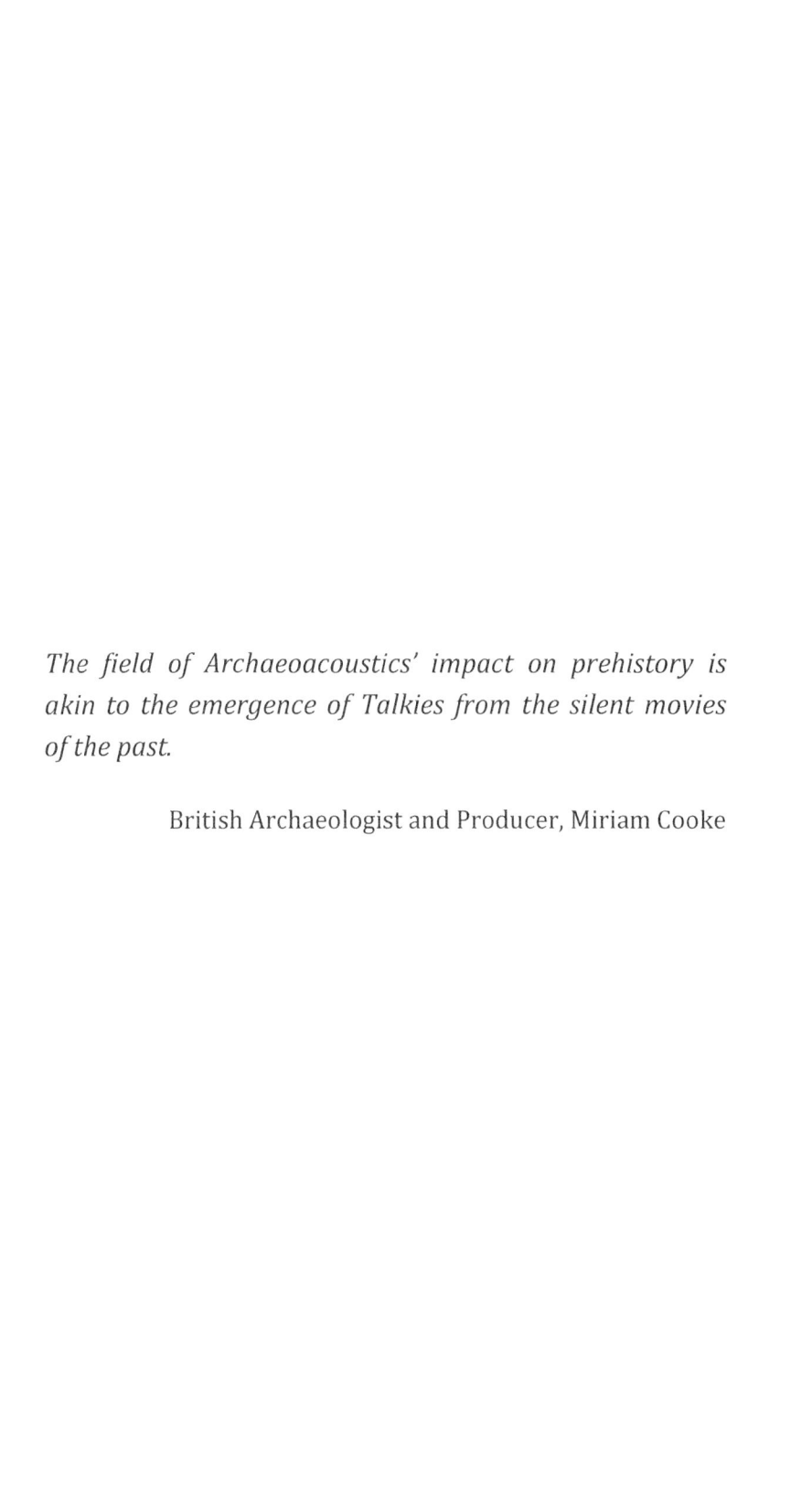

The field of Archaeoacoustics' impact on prehistory is akin to the emergence of Talkies from the silent movies of the past.

British Archaeologist and Producer, Miriam Cooke

Chapter 3

Orientation

3.1 The Sapient Development Timeline

How comforting to think that earth's present population is the product of survivors. (At least they lived long enough to reproduce.) Everyone reading this text today represents the living end of a long line of folks who have existed since human beings evolved from earlier creatures.

It is thought that people who looked very much like us and had brains like ours evolved about 200,000 years ago. Early populations figured out how to master fire, which made a big difference in lifestyle. Cooking food prefaced important changes. As reported by Rebecca Boyle in a compelling feature article for *Popular Science*, "Eating cooked food allowed these early hominids to spend less time gnawing on raw material and digesting it, providing time—and energy—to do other things instead, like socialize. The strenuous cognitive demands of communicating and socializing forced human ancestors to develop more powerful brains, which required more calories- calories that cooked food provided. Cooking, in other words, allowed us to become human."[15]

Megaliths, Music, and the Mind
Linda C. Eneix
Copyright © 2024 Jenny Stanford Publishing Pte. Ltd.
ISBN 978-981-5129-25-0 (Hardcover), 978-1-003-46823-3 (eBook)
www.jennystanford.com

It took time to learn how to talk to each other and to invent language. We also had to figure out how to make clothing and identify what things were safe to eat. As humankind developed, people learned how to tell directions and forecast weather from the sky. They figured out what to do about sickness and injuries and even dental cavities.

Scientists believe that around 40,000 years ago, something triggered the human brain to develop abstract thinking. With that, people were able to picture in their minds something that was not immediately in front of them. Figurative art was invented. The prehistoric "venuses" go back 25,000 to 40,000 years ago, as does the oldest known statue, known as "Lion Man".

It may have been earlier, but 16,000 years ago seems to be the point where that art was expressed in pigment as cave paintings such as the famous murals at Lascaux. Surely, the artists did not have one of these beasts posing in front of them and holding still as the painting was being done. Imagine them painting the stone walls and ceilings by the light of little stone lamps filled with animal fat and moss. It's unlikely that they were doing so just because they were bored. There was some sort of serious ritual happening here. (Incidentally, something else interesting happened around this time. Dogs attached themselves to humans and became pets.)

It has been observed that in the famous Paleolithic caves of France and Spain, the beasts are painted in the parts of the caves where the echo is the strongest. Already, it seems, there may have been an association between the visual countenance of the animal and the audio boost that the environment gave to a human imitation of its fierce grunting and bellowing.

Up until the last Ice Age ended around 12,000 years ago, everyone on the planet lived much the same way. They moved around in tribes or family groups that helped each other stay fed and clothed and sheltered. The good of the group was more important than any one individual and there was little, if any, sense of personal property. In fact, there was little property. They likely decorated themselves. Every native tribe we know of has done that. We are still doing that!

Many of them definitely visited caves, but contrary to popular belief, most were not permanent "cavemen". Caves don't exist everywhere and they are not the easiest places in which to live. Far more often, housing was made out of what there was to find near the food sources, like trees and brush. Each time they camped for a while, setting up near drinkable water would have been a top priority. Whatever it was that they constructed would be left behind when the group moved on, because wagons hadn't been invented yet and pack animals had not been tamed. Everything had to be carried when the tribe moved to follow game or before a change of seasons. Basically, as with nomads everywhere, once they had moved on again there was little trace of human impact that would last very long.

Then came the notable activity of about 12,000 years ago that would change the world and the course of human development forever. In an area at the eastern side of the Mediterranean called "The Fertile Crescent", in the Asia Minor part of the world that in ancient times was called Anatolia, some of those people decided to build something monumental. There would be no turning back.

In 1995, in what is now southeastern Turkey, not far from the Syrian border, German archaeologist Klaus Schmidt began uncovering the ruins of a site that would revolutionize what we thought we knew about the ancient world. When the carbon dates came in, the archaeological community was shaken. Everything we thought we knew about the sequence of human development and the advent of civilization was wrong. For some reason, nearly 12,000 years ago, what was likely a collection of tribes or family groups of people began serious construction projects.

Figure 3.1 (*continued*)

The Sapient Development Timeline | 55

Figures 3.1 Göbekli Tepe excavation site, Southeastern Turkey/Anatolia, ca. 11,500 years old.

56 | *Orientation*

Figure 3.2 The above three images are of the Göbekli Tepe excavation site, Southeastern Turkey/Anatolia, ca. 11,500 years old.

The stone circles that are being uncovered at the site called Göbekli Tepe have been identified as the oldest buildings in the world—period. These are full-blown monumental enclosures. The authorities say there are hundreds of them in the region, still buried under the soil and rock.

The unexpected news that so shook the scholars is that people were building and using shrines like Göbekli Tepe before they started farming. Parts of this site are dated as far back as what is called the pre-pottery Neolithic. Pottery is a very telling clue. The reason people didn't make pots before this time was not because they didn't know how. It was because they didn't want them. What hunter/gatherer person who believed she would be moving on when the food ran out would be interested in making heavy clay vessels? That would change when people began settling in the same place for extended periods, and that would imply some agriculture.

There is no evidence of any village or settlement near Göbekli Tepe that dates to the time of the monument building—nothing like foundations or remnants of walls or permanent hearths. There is no nearby source of fresh water. Nobody lived out their lives here when the monuments were raised. At the time of the building of the Göbekli Tepe sanctuaries, people came and went, most likely on a seasonal basis.

And yet—at the time of first construction, large groups would be collected for the project. While the workers with their big muscles were building and decorating the great stone meeting centers, they were not hunting. Of course, everyone still needed to eat.

As far as an acoustic awareness at this time is concerned, elements of the original construction design at Göbekli Tepe are suggestive. Of course, it's impossible to measure the resonant behavior of the sanctuaries today, but if these stone enclosures were roofed when they were in use—and we think they were—the sound quality inside would have been very different from what was experienced under the sky or in brush shelters. To the people of the time, who had never been inside any large building and lived most of their lives outdoors, such a thing in itself had to be a marvel.

3.2 Neolithic Revolution

On the plains surrounding Göbekli Tepe, it would not be long before agriculture took hold. Here, food animals were domesticated, grain crops were cultivated. It is as if the intentional production of food began as a way to support the community efforts of building and using the stone enclosures: feeding the workforce, providing for large gatherings for an extended period of time. This is a bit simplistic, but the carbon dating is clear. Scientists now think that this was the reason for one of the greatest changes humanity has ever made. Knowing how to farm and control their food supply was a radical advancement. It was not an overnight shift, but it was so contrary to what had been a universal existence everywhere on the planet, that one can't help wondering about the trigger.

Dr. Schmidt, the man who first identified the Göbekli Tepe site, always maintained that extensive, coordinated effort to build the stone enclosures literally laid the groundwork for the development of complex societies.

"This shows sociocultural changes come first, agriculture comes later," says Stanford University archaeologist Ian Hodder, who excavated Çatalhöyük, a prehistoric settlement not far from Göbekli Tepe. "You can make a good case this area is the real origin of complex Neolithic societies."[16]

Thus began the "Neolithic Revolution". (Neolithic is a word that is mainly used to describe a time period, which in the West began after the last Ice Age and continued until somewhere

Neolithic Revolution | 59

Figure 3.3 Illustrations of area around Göbekli Tepe, ca. 11,500 years ago.

Figure 3.4 Çatalhöyük, ca. 9,500 to 8,400 years old.

around 4,500 to 5,000 years ago, depending on the location.) Not confined to the Göbekli Tepe region, it caught on over time and spread outward in all directions.

It is not easy to condense the impact of this change in lifestyle. Once people were farming and knew where their food was coming from they could remain living in one place. That meant they could build sturdier houses and start collecting things. Without the need to be mobile, the next thing you know it's more permanent furniture and heavy looms for weaving larger textiles. Decorating and developing fine ceramic pottery makes sense when one doesn't need to make the containers easily portable. They created ever more sophisticated tools that gave them an edge over the tribes who still spent most of their lives chasing and collecting their food.

As people prospered, farming settlements became towns and those became cities and civilization was born, consistently building on the inventions of people who came before. There would not be such an impactful development again until the industrial revolution brought automated machinery, and that could never have happened without the steps that came first. None of the technology and lifestyle that we have today would be possible without agriculture and the drive of those first farmers. We can see how the mystery of what prompted the building of the megalithic monuments is the heart of the "Sapient Paradox".

Over time, in new settlements that rose to the east, the wheel and metal-working were invented. Techniques were improved and new inventions could be tried. Surpluses prompted the development of trade. A bit later, a system of symbols was invented to keep track of transactions long before an alphabet came along to document conquests. Unfortunately, disputes over ownership of goods and territory also became part of life.

Yet people thrived and families grew in this part of the world. As time progressed, they moved great distances and launched the societies that wrote history.

3.3 Genetics and Migrations

The Çatalhöyük excavations, located about 300 miles from Göbekli Tepe and dated to several thousand years after, reveal an extensive settlement of the Neolithic period. There is an abundance of informative material about lifestyle and beliefs from that site, although so far archaeologists have found no corresponding monuments.

We can also learn quite a bit about the Neolithic scenario by studying the museum collections and reports from Malta. We have mentioned the treasure trove there that is nearly as valuable as if people from the Stone Age had left us a time capsule. The only thing missing from the Maltese collection is a surviving domestic building. It should be noted that archaeologists have found traces of mudbrick dating back to the period: an unusual choice of building material for a rocky island. Could this be a building habit brought from a middle eastern homeland, perhaps?

The material culture of the people who lived at Çatalhöyük from about 9,500 to 8,400 years ago and those who later settled on Malta is close enough to invite speculation about a link of cultural descent between them. It is fascinating to consider that from Turkey to Malta to Portugal, Spain, France, Scotland and Ireland, everywhere a Neolithic Culture is identified, at least some of the same materials are present. The "Neolithic Kit" includes (but is not limited to) red ochre, obsidian (volcanic glass), flint, polished stone axe heads, spindle whorls, beads, incised pottery ware, goats, bread wheat, barley, fat lady sculptures and megaliths. Domestic animals usually also include cattle, sheep, and pigs. Barley, by the way, nearly always goes with the making of beer. They found basins for making beer at Göbekli Tepe. There are great stone vats in the Neolithic ruins of Malta. There are stone basins inside the passage "tombs" at Newgrange and Cairn L, Knowth. Some human things will never change. (Maybe beer had something to do with why farming was invented.)

For a long time there was debate in the scientific world as to whether it was ideas or actual people who spread the farming lifestyle. These days, with our advancements in genetics, we have the answer to that.

Fine-tuning the dates places major world-changing population movement at somewhere around 8,500 years ago. It looks like there were additional smaller waves that followed later. In a fascinating parallel of about the same timing, Dr. Schmidt reports, "During the Neolithic, and for reasons unknown to us, settlement refuse was deliberately dumped into Göbekli Tepe's megalithic architecture which, as a result, was sealed and protected until its discovery in the mid-1990's." One can almost imagine a ritual "closing" of the sanctuaries in the same way that we might shutter a house that we know is going to be vacant for a while.

Maybe these large-scale movements were because the population got too big. There is a very old folktale of drought and an Anatolian king who divided the population: half to stay and half to go out into the world in search of better conditions. This might actually carry some weight. The people who would have stayed in the area between the Tigris and Euphrates likely developed *Mesopotamia*, which is itself a word derived later in time from the ancient Greek for "between the two rivers". When one starts considering Babylonians and Sumerians and Akkadians, the cultural territory is more familiar. We don't know details yet, but the science of Genetics supports the theory. Of course, people don't stay in the same place indefinitely and they don't keep moving exclusively in a single direction. They intermarry and go backwards and confound the nice neat map of heredity that we would like to see.

What is clear is that eventually the people from this area were the ones to take the idea of farming into Europe. Apparently, it was largely a peaceful transition with easy assimilation of existing populations. The genetics are compelling with both Y-haplogroup and mitochondrial components. The agrarian lifestyle must have had obvious attractions.

Independent studies have confirmed that the bread wheat and barley in Europe (and later in America and beyond) is descended from plants that once grew very close to Göbekli Tepe. Crops like lentils, chickpeas and flax were first grown near there. Specialists have traced the DNA of farm cattle and domesticated goats in Europe right back to this same area between the Tigris and Euphrates rivers. They can identify the trails of the spread of goats from Anatolia, beginning at about 8,800 to 8,500 years ago, via two routes: The Mediterranean and the Danubian.

It seems that cattle became livestock at about the same time, from the same area. It has been reported that about 1.3 billion of the world's taurine cattle are descended from a small herd of about 80 ancestors domesticated from local wild ox (aurochs) in ancient Anatolia, not far from the Göbekli Tepe area. The author of the report shares an interesting detail: "After an initial breeding phase lasting some 1.5 millennia in an area between the Levant, central Anatolia and western Iran, domestic cattle started to appear in western Anatolia and southeastern Europe by 8,800 cal. BP, southern Italy by 8,500 cal. BP, and Central Europe by 8,000 cal. BP. Archaeozoological residual lipid and lactase persistence data point to an increasing economic importance of cattle for meat and milk production."[17]

Writing for BBC.com, Jason G Goldman explains one example of a cultural tradition that has resulted in genetic changes that can be inherited biologically from one generation to the next. "You shouldn't be able to drink milk. Your ancestors couldn't. It is only in the last 9,000 years that human adults have gained that ability without becoming ill. Children could manage it, but it was only when we turned to dairy farming that adults acquired the ability to properly digest milk. It turns out that cultures with a history of dairy farming and milk drinking have a much higher frequency of lactose tolerance—and its associated gene—than those who don't."

And where did this genetic change happen? Anatolia. Who is responsible for it? The same people who used Göbekli Tepe. The first farmers who domesticated cattle. The ancestors of the Neolithic settlers who moved across the Mediterranean and Europe. So if you are an adult enjoying dairy products, you can thank those first architects; the ones who apparently were exposing themselves to particular sound behavior.

Wenzu the Bull is one of the last living direct descendants of the Stone Age cattle that were brought by settlers to Malta around 6,000 years ago. Isolation on the islands prevented later assimilation with other bovine varieties and kept the genetic line pure. He bears a striking resemblance to a carving of his ancestor made in one of Malta's prehistoric monuments.

The path of livestock movement is matched by the route of Neolithic figurines and painted pottery. The same time period sees the spread from Anatolia of the Indo-European language along the same trail of Western migration. This latter includes a further split 7,000 years ago: one arm resulting in the ancestor of Italic, Celtic and Germanic language, while the other is believed to have become the ancestor of Tocharian, which ended up far to the East. An article in "New Scientist" elaborates: "An analysis of related words in 161 languages

66 | *Orientation*

This column from Gobekli Tepe may illustrate the casting of a weighted net to capture a wild sheep.

Making yoghurt the traditional way.

Figure 3.5

Genetics and Migrations | 67

Figure 3.6 Carving at Tarxien Temple, Malta, ca. 5,200 years old.

suggests their shared roots lie in the Middle East—a conclusion that also fits with DNA evidence. The common ancestor of Indo-European languages, which are now spoken by close to half the world's population, was spoken in the eastern Mediterranean around 8000 years ago, according to an analysis of related words.

"Indo-European languages, spanning from English to Sanskrit, have long been thought to share a common ancestor. The first linguist to make this link, William Jones, said in a lecture in 1786 that no linguist could examine Greek, Latin and Sanskrit together 'without believing them to have sprung' from some common ancestor."[18]

An odd coincidence worth mentioning is a unique type of archaic polyphonic singing that has been preserved by shepherds in remote regions of Corsica, Sardinia, Albania, Istria, Bulgaria, Georgia and Anatolia. The pattern of distribution makes it tempting to wonder about the survival of a custom from a common ancient source, carried by the Neolithic migrations that went out from Anatolia in the Stone Age with their traditions, their animals and their agricultural lifestyle.

Another clue to identifying this trail is the occurrence of blue, green or grey eye color. Based on research conducted by a team at the University of Copenhagen, it is now widely reported that the mutation resulting in blue eyes occurred for the first time around 10,000 years ago in the Black Sea area. Related reports are fascinating, as it is carried in the mitochondrial (from the mother) DNA. Blue eyes are depicted in artifacts from ancient Egypt and have been identified from DNA in 7,000-year-old remains from Northwest Spain. (One could also find some interesting material about the genetic origins of blonde hair color if interested in these things.)

A re-evaluation of ancient DNA recovered from Ötzi the Tyrolean Iceman, the 5,300-year-old glacial mummy recovered from the Ötzal Alps, indicated that about 90 percent of Ötzi's genetic heritage comes from Neolithic farmers from Anatolia. Phenotypic analysis also revealed that "the Iceman likely had darker skin than present-day Europeans and carried risk alleles associated with male-pattern baldness, type 2 diabetes, and obesity-related metabolic syndrome."[19]

Just as Judaism, Zoroastrianism and Hinduism have preserved practices and concepts that are still in use nearly 4,000 years later, a cultural tradition appears to have traveled with people and persisted over a long time. At least some of those pioneers in both paths of prehistoric "European" migration took with them a cultural tradition of building with huge stones. The same material would not have been available everywhere and there would naturally be variation. But examples of the places of worship around the world, even in their very wide differences, still illustrate a recognizable constant within their particular belief systems. How much of that megalithic tradition had to do with sound? How much was on purpose and how much was by accident? Maybe we will find out.

Malta's temples and Göbekli Tepe are pre-planned structures that completely enclose interior space. While still meeting the criteria for a megalithic monument, Stonehenge is not the same sort of approach because it never had walls or ceilings. It could have been a local interpretation to serve a purpose with available materials as styles changed over time.

A science editor for the BBC reported, "The ancestors of the people who built Stonehenge travelled west across the Mediterranean before reaching Britain. Researchers compared DNA extracted from Neolithic human remains found across Britain with that of people alive at the same time in Europe. The Neolithic inhabitants were descended from populations originating in Anatolia (modern Turkey) that moved to Iberia before heading north. They reached Britain in about 4,000 BC."[20] There is an abundance of material available to researchers these days about movements into Ireland, Scandinavia and elsewhere. Apparently, it was a peaceful transition and the agrarian lifestyle took off like wildfire in these areas as it did elsewhere.

We know something about the methods of the Phoenicians who were sailing the Mediterranean in the Bronze Age, but almost nothing about maritime movements earlier than that. And yet—people got over mountains, through vast forests and across deep water to islands somehow. They had their animals, including oxen, with them! They took seeds and plants and revolutionized the land everywhere they went with the idea of agriculture. Much later in time, subsequent generations of their livestock and seeds would cross the Atlantic with the first settlers in the New Lands.

Colonization of the Americas, Australia, New Zealand, South Africa and beyond drew from this gene pool. Immigrations in historic times are well documented. Those first farmers of the Göbekli Tepe area were the ancestors of around a third of the people on the planet today. There is a lot to think about there. For better or worse, the whole 21st century world has come to be impacted by developments of people who descended from these ancient farmers and their ancestors in Anatolia. They could be said to have parented Western civilization. The start of agriculture and the advancements that followed were a bit different in Asia, as they were in Central and South America. But in our time, even in remote locations of the world, they know about internet, McDonald's and Coca-Cola; most people have seen a Hollywood movie or at least heard of Superman and Mickey Mouse.

Where is the life we have lost in living?
Where is the wisdom we have lost in knowledge?
Where is the knowledge we have lost in information?

T. S. Eliot

Chapter 4

Megaliths

4.1 The Earliest

The stone circles of Göbekli Tepe have also collectively been called the oldest temples in the world. Four monumental enclosures with 51 pillars in situ have so far been excavated in layer III. Ground-penetrating radar shows at least 20 more enclosures buried under the soil on just this one hill.

There had to be some trial and error on the way to these huge works that have survived for thousands of years. Maybe the earliest shrines were mud or wood and just didn't endure. The likelihood is that they were cleaned up and remade in the same spot with better technique as the ancient people worked out better engineering. Although we don't have the whole story yet, it is emerging.

"Today we know a lot about this pivotal period," writes Schmidt. "Several sites have been excavated like Çayönü, Nevali Çori, Hallan Çemi, ot Körtik Tepe in Turkey, Nemrik and Qermez Dere in Northern Iraq and Mureybet, Jerf el Ahmar, Tell Abr, and Tell Qaramel in Northern Syria. But no other site

---

Megaliths, Music, and the Mind
Linda C. Eneix
Copyright © 2024 Jenny Stanford Publishing Pte. Ltd.
ISBN 978-981-5129-25-0 (Hardcover), 978-1-003-46823-3 (eBook)
www.jennystanford.com

of this period yields all the elements discovered at Göbekli Tepe: it is not only unique in its location on top of a huge hill; its monumental and megalithic architecture has not been found at any other site of the Near East."

Described by Dr. Schmidt, "The main features are the T-shaped monolithic pillars. They stand in a circle, looking toward a pair of pillars in the center." Remarkably, much of it looks similar to the Maltese temples which would come thousands of years later and half a sea away. Although they are revealed now in excavation, the shrines at Göbekli Tepe were never freestanding. The stone walls have always been supported by the earth that surrounds them.

The columns at Göbekli Tepe have been carved in bas relief. They are heavily covered with beasts and creatures. The animals are male, where it shows. They include scorpions and snakes and birds of prey: many scary creatures, showing their teeth. There are quite a few that look like foxes or possibly dogs. If the latter, perhaps the dogs were trained for the hunt. One of the side columns in enclosure D seems to be an illustration of technique, saying that if you throw a weighted net over a wild sheep, the odds are good that you will be cooking soon.

Toward interpretive guidance, one of the most important Native American beliefs is that everything from stones to plants and animals to birds has a unique spirit. Native Americans believe that the animal spirits serve as human guides. They hold that the spirits appear in human dreams to show the path and serve as their guardians. We can't know what Göbekli Tepe's artisans intended but these carvings were far more than just decorations. They must have held some sacred meaning, mystical in design and illustrating a cosmology that we are never going to fully understand.

The Earliest | 75

Figure 4.1

76 | *Megaliths*

Figure 4.2

The Earliest | 77

Figure 4.3

If your life was sustained by chasing and gathering wild things and you believed that nature gave spirit to the beasts that you hunted, how would you express that? If you wanted to communicate your wishes and desires to that spirit, how would you do it? If you believed that power of life and creation lay in the earth that directly and exclusively provided for all your needs, and if you believed that such a power could communicate with you, how would you express that? Perhaps that is what we are seeing here in a prehistoric expression.

Karahantepe, a new discovery near Göbekli Tepe, was made by the same culture but possibly even a bit earlier. Still under excavation, the Karahantepe complex includes what's left of two chambers, connected by a hole in the stone. One chamber is a large oval hall that would accommodate a good sized group for ritual activity.

On the other side of the hole, there is a smaller room carved out of the bedrock, about 6½ feet deep. The suggestion has been launched by a popular writer that this room, about 20 feet in diameter, was a place where initiates or people looking for connection or communication with some kind of otherworldly force or influence or deity would come to do their attunement in altered states of consciousness. Ten of the 11 columns were sculpted out of the solid bedrock as the space was chipped out. It is claimed by workers on the site that the 11th may have functioned as sort of prehistoric tuning fork. So far, there is nothing official published about that statement. Certainly, when it is completed, an archaeoacoustic evaluation will be revealing.

Now, if one wanted to visually represent a voice calling out from the earth, this would do it. The face is more than three times as big as a human's and at about the same height from the floor, in perfect placement for a conversation.

The Earliest | 79

Figure 4.4

The stone shrines like Karahantepe and Göbekli Tepe were not dreamed up piecemeal as the builders went along. They are the earliest example we have of monumental architecture. The key to this identification is that they were carefully designed and planned in advance. It was necessary to know exactly where each stone was going to be placed before it was moved from where it was quarried because moving a megalith around is not like rearranging furniture. There had to be leadership and good communication to get one of these enclosures put together.

Standing at Göbekli Tepe, one marvels at the isolation of the place. Sound would have travelled a long way across those plains in the middle of nowhere. An official call-to-gather would be hard to miss in the open and mostly silent landscape. We might ask: for what purpose would they gather?

Most of the indigenous or native people that we know about get together with other groups from time to time to share news, see old friends, and meet fresh blood for strengthening the biological health of the tribes. It may have started small and grown huge as the word spread.

Authorities say there are at least 11 more hill sites out here that haven't even been touched by archaeologists yet. The building of the shrines was certainly not a flash in the pan. There was some serious motivation at work. Some might say that it looks like this whole area was a destination. Imagine trekking across the plains for days or even weeks, and then seeing these shrines appearing in clusters on the horizon—when you have never seen a big building before in your life. In its heyday, it would have been like a prehistoric Disneyland here.

Were the stone circles roofed? Dr. Lee Clare, the director of the excavation project at Göbekli Tepe, thinks that they were. He points out features that support a convincing argument, including the care that was taken to regulate and coordinate the height of the interior columns. The sound quality inside them would be altered with enclosure by roofing, although we can't say exactly how without knowing something about the material.

As suggested in the introduction, hunter/gatherer tribes who had been conducting special activities (with or without paintings) in revered caves, might have come to understand that when hunting takes you where there is no cave, you could gather some stones and build one yourself. It is now clear that ancient builders did produce sound phenomena that were outside of normal in their monumental stone structures. There are plenty of possibilities that are not so clear.

Did somebody watching ants work together suddenly get an idea? Could a person who has known no other life than living in harmony with nature suddenly consciously decide to change the land as it occurred naturally? Was there one person who rose up, or a tradition that took on extra importance? Was it competition for food and a desire for "claiming" territory? Why do it with stone? Were megaliths perceived as the "bones of Mother Earth"? Did the builders engineer special features with the intention of making an acoustic environment? Was it pursuit of such sound phenomena that altered the course of human development? Clues for addressing some of these and other questions may come from an architectural evaluation. At the very least, we can better understand a few things about the capabilities of very human people in the earliest days of transition.

We can start with everyone's first question: how did people with nothing but primitive tools move the stones? With all our scientific and engineering know-how thrown at this question, there is no agreement. We are asking it, however, as modern people who know about cranes and heavy machinery. We are naturally thinking about how we would approach such a thing today. Eleven thousand years ago, folks didn't see it as a nearly impossible task, so they found a way to do it. The proof is standing there in front of us.

There is a fun story about a megalith that had fallen at one of the Maltese sites in an area where archaeologists wanted to work. After some discussion, the supervisors decided to get the Royal Engineers to come the next day with equipment to put it back up to the obvious place from which it had fallen. Three Maltese men were chuckling on the sidelines listening to this conversation. These workmen had been assigned temporarily from their jobs in the quarries. They knew the local stone and were well-seasoned about what was the best way to maneuver it. The excavation team arrived the next morning to discover that the workmen had already put the stone back in place 1.5 meters above ground level, using a plank and brute strength! It had taken them about half an hour.

These places were made before the invention of the wheel. In antiquity, sledges, levers, a lot of sweat and a profound understanding of the material are all likely factors. We must add ingenuity, teamwork and confidence.

Round stones like giant ball bearings are found around Malta's temples. At the Hagar Qim site, there can be seen sockets cut into the bottoms of several exterior uprights so that they could be pivoted into position. Convincing evidence can also be seen at the Tarxien Temple, where one of these stone roller balls has been left exposed in position between the bedrock and a thick paving stone.

As they take pencil to paper these days, (or pointer to screen,) architects and planners for public buildings seem to be first concerned with factors like traffic patterns and available parking. How did the ancient builders decide where to start? Sites in prehistory were not chosen at random, although they were probably based on something more esoteric.

4.2 Placement

With regard to the original builders on Malta, architect Richard England explains, "The buildings they produced related to both earth features and the heavens above and provide manifest examples of both complex building techniques and spatial layouts. The locations chosen were also always places of power and energy, sites identified for man's awakening through the somatic and spiritual qualities of the site itself. These people seemed to be capable of the reading of earth's forces in a manner not dissimilar to Eastern geomantic practice of 'Feng Shui'."

One possible factor in site choice might have been infrasound: sound that is below the "normal" limit of human hearing. Sound, even at infrasonic levels, can still be felt even if not heard. This frequency range is utilized for monitoring earthquakes, and also to study the mechanics of the heart.

For us, vibration is felt in the skin via Meissner's sensors or mechanoreceptors, which are very common in the palm of the hand and in the chest. Soothsayers are said to have them very highly tuned, according to Dr. Paolo Debertolis, who is an avid researcher in the field of Archaeoacoustics. "We detect infrasound all over in ancient sites," he says. Lending weight to a proposition that early man had better "detectors" than we do, he describes a recent investigation: "Something really wild was discovered at a site in southeastern Turkey near the Syrian border."

Sogmatar, in southeastern Turkey, was an important cult center in which the people of Harran worshipped the moon and planetary gods during the time of the Abgar Kingdom around 2,000 years ago. The semi-subterranean temple examined by Debertolis belonged to the god Sin, god of the moon. Inside, an audible resonance frequency of around 93 Hz was found;

within the range that could be expected. Then Debertolis picked up the infrasound. "With silence it is very easy to record a strong natural frequency coming from below the soil of around 14 Hz which was also found outside this cave. The natural infrasound vibration is stronger inside the temple than in open air. This is normal because the temple acts like a resonance box for the people present in this cave without the interference of outside air movement."

It wasn't until later in the lab that Debertolis and his team found something more mysterious. "For testing the movement of the air molecules, a large number of photos and video of the main and secondary room were taken using an ultraviolet camera. Being inside a closed room with low natural light, it was unnecessary to use filters to stop infrared rays affecting the images. After the analysis in the PIV vector program, a strong spiral magnetic field on the wall located on the right of the central niche was observed. Here are carved several relief figures of their gods. In the center of this spiral there is a total quiet as in (the eye of) a hurricane. There is no explanation for a magnetic field of this shape and behavior."

We don't know how the locations for ancient shrines were chosen. We don't know how a divining rod works. It does seem plausible that people built in a particular place because somebody felt something, whether or not they could explain it.

Prof. England addresses a sort of *feeling* thus: "There is no doubt that architecture has a particular voice for different spaces and that different materials reverberate differently. It was Adrian Stokes who said that 'a long sound with its echo brings consummation to the stone'. Architecture remains mood manipulative not only through its visual power but also as an instrument of sonic manipulation. Have we not all experienced spaces to lower one's voice, spaces to linger in or spaces which encourage meditation?"

For site choice, add to all these noble factors the proximity of natural outcroppings of appropriate stone, which would be high on the list. Sightlines and views also count in the overall scheme of things. And we must consider the possible availability of fresh water, even if it had to be carried to cisterns for storage. Rainfall could be collected there, too.

4.3 Design

At the time of this writing, Göbekli Tepe and its sister sites are the oldest known monumental buildings on the planet. The exposed built walls, as mentioned earlier, have always been supported by surrounding soil and fill. Malta's temples keep the title of the world's oldest freestanding buildings. They exhibit the earliest known use of corbelling, forecourts, and ring compression systems that are stunning in their engineering sophistication.

Although Göbekli Tepe and the Maltese megalithic temples are separated by thousands of years and many miles, for this purpose we can view them together as structures that represent the birth of architecture, far ahead of anything else so monumental. The symmetry of their arrangements and the grace of scale are identical. Both demonstrate plastering on the interiors, and elegant floors finished in the same method.

Dr. Schmidt estimates that for the construction of one of the megalithic enclosures at Göbekli Tepe, several hundred people must have gathered for many weeks; "Without doubt, one can expect that these meetings were arranged as extended feastings." He notes that the dimensions of the megalithic stones of Göbekli Tepe are similar to those at Stonehenge, although 6,000 years divide the creation of the monuments. "No doubt the amount of time, energy, craftsmanship, and man-power necessary for the construction and maintenance of this Upper

Mesopotamian site," he continues, "is indicative of a complex, hierarchical social organization and a division of labor involving large numbers of individuals at the site."

At Göbekli Tepe and in Malta, the workforce and motivation were probably about the same, and put the same pressures on their communities. The initiative of an island population that was starting from scratch in Malta in many ways corresponds to the forerunning builders of Anatolia. Examining the abundant and remarkably preserved archaeological remains of both locations helps us begin to address the bigger question. For what possible reason would humankind start doing this kind of major enterprise after thousands of years of building nothing more than utilitarian shelter? A logical driving force would seem to be spiritual concepts shaping what Dr. Schmidt calls a materialization of their complex immaterial world. In other words, it was for ancient gods.

There is disagreement about how the structures were roofed. Whether in stone or with timber, brush and skins, there are arguments for both, as well as a faction that claims they were always open to the sky. The author believes that we can discard this last idea. The evidence that argues against it for both the Maltese monuments and for much older Göbekli Tepe is that the bas relief carvings in the soft limestone that comprises the ancient sites in both locations would never have survived extended exposure to sun and rain. Deterioration in the Malta sites can be seen clearly by comparing exposed decorated surfaces today to photos taken when the stones were first cleared of the debris and soil that had been shielding them. Unfortunately, protective covers did not get into position in Malta before much detail had already been lost. Someone had the foresight, however, to put many of the most highly decorated stones into the national museum and replace them on site with replicas. Most of these saved stones came from a temple that was fully cleared only a hundred years ago.

Design | 87

Figure 4.5

88 | *Megaliths*

Figure 4.6 Note that the animals carved in the Malta temples were no longer wild and mean, but were of the domesticated variety.

In the case of the Maltese complexes, the conservation covers have noticeably changed the quality of the sound beneath them. There are echoes now under the modern membranes, even with open sides for airflow. It's a good guess that the monuments as originally roofed and sealed were even more acoustically impressive. With their concave façades suggesting an external stage, one wonders if sound from the courtyard traveled across the water to boats at sea. Perhaps the temple itself was used as an instrument, like some giant siren.

As illustrated, the layouts of the Malta sites are a complex expansion of Göbekli Tepe's pattern of a singular round or oval enclosure. In the Malta temple structures there are drillings for hinges at the portals. Doors have not survived, but were most likely heavy wooden ones. The jams and sills were configured to open inwards. Sightlines are deliberately controlled and restricted in some areas. Several of them exhibit special window-like openings for communication between outside and inside.

The interiors of the monuments were initially designed with a central corridor, running from the formal entrance to an apse at the back. Lobed chambers spread out on either side of this corridor. Curves and more curves—even the facades gently arc to hug large paved or leveled forecourts where it is presumed the ancient communities would gather. The set-up would have made a perfect backdrop for performance.

Portals defining the central corridors are frequently described as post-and-lintel trilithons: two standing megaliths on either side of the opening, capped by a horizontal slab at the top. A closer look shows a fourth stone: another horizontal slab that lies across the floor. These threshold slabs stabilize the upright post stones and keep them from kicking inward at the bottom like "pigeon toes". Although the steps that they produce are shallow in height, they also serve to trigger in the visitor a conscious transition from one space to another.

The threshold paver stone chosen for the entrance to the southern temple at the Mnajdra complex has a unique lateral vein running through it, suggesting that it was quite a prize find. Stepping over it and through the stone enclosed "gateway" passage of about five or six feet in depth, one moves into a great oval void: the midst of the matched semicircular rooms to either side. It also becomes clear that there are more rooms lying beyond the next portal arrangement.

Large stones stand side by side to define the walls of the semicircular apses. They have been angled very slightly at the base of the wall to press into a fill of soil and broken pottery that was packed between the walls of the chambers. The ring of corbelled stones that make the upper portion of the walls butts against the top portions of the portal post stones and make a cap above the larger upright slabs. The great lintel stones that span the top of the portal posts serve to space the two half-circles apart so that their inward thrusts cancel each other out. Like the beginning of a dome, additional courses in each apse extend slightly over each lower course as they go up, thereby reducing the span across the chamber. Two of the temples demonstrate in the surviving topmost level of these stones that they tilt slightly down and inwards, instead of throwing all the thrust into the walls. When it was completed, the whole system locked itself into perpetuity. It's an amazing feat of engineering that we could certainly call *advanced*.

In Malta's Hypogeum, features in many of the chambers are carved to represent the megalithic interiors of the temples above ground. They have their ceilings carved to imitate a built roof of corbelled masonry, with one course overhanging the one below to narrow the span as it goes up. This is a clear architectural indicator that the temples had a roofing system, and that would certainly have impacted on sound quality.

The know-how for something like this does not just pop into being. There are none of the expected prototypes to be found in Sicily, the nearest land. These buildings were begun and perfected in Malta. Could there have been some sort of cultural memory that arrived with the first people?

Several factors offer intriguing parallels. The first is the curved wall, the acoustic effectiveness of which is demonstrated in today's theatres and concert halls. We design them this way on purpose to accommodate the hearing needs of an audience. In the larger and older structures like Göbekli Tepe and the temples on Malta, the whole interior chamber is rounded. Frequently, there are low stones around the perimeter that would be suitable for seating a small audience.

The Greeks used the horizontal arcs in their theatres for the same reason, putting the audience inside the arc with the stage at one end instead of in the center. In "passage graves", tombs and megalithic structures elsewhere, the wide curve has been used on the exterior, reduced to a stone arm-like façade embracing the approach to the entrance. This feature can be found in the younger megalithic ruins of Sicily, Sardinia, Menorca, Portugal, the British Isles and beyond. It is a good ceremonial monumental form. If it was part of a prehistoric acoustic plan, the external curved walls would have been the reflector of sound to an audience that was outside in the forecourt.

Notable, but absent from the reports is the presence at Mnajdra South Temple in Malta of a small elongated conical stone fitted into the base wall at the entrance. Such stones are commonly found in the Babylonian materials and have been associated with commemorative placement in important buildings of the Mesopotamian Middle East. In 2012 one of these which was inscribed was seized at London's Heathrow Airport and returned to Iraq from where it had been taken illegally. The inscription promises a curse on anyone who would destroy it.[21]

92 | *Megaliths*

Figure 4.7 Megalithic structures in Turkey, Malta and France.

Design | 93

Figure 4.8 Megalithic structures in England, Sardinia and Russia.

The inward-opening doors of Malta's monuments are placed to keep things out, not inside. The buildings are engineered to be looked into, with sightlines restricted. This goes along with other evidence in the older monuments which points to restricted use of the interiors by an "elite" or privileged few. Even the "communication" portholes that appear in several of the Malta temples are cut through the stone in such a way that not much of the inside can be seen through them without contortions. These are also sometimes referred to as "oracle holes" although that is a modern naming for them, colored by what someone thought they might be.

Interesting perforations in the stone also show up at Göbekli Tepe. Klaus Schmidt pointed out that "Similar objects are well known, e.g., in the megalithic barrows of Corsica and Atlantic Europe. Stone slabs with a central hole had been placed in several barrows vertically in that way, so that the stones define the entrance leading into the darkness of the grave."

It needs a very different way of looking at all the places of antiquity to consider this concurrence in depth. Since the tradition of Archaeology is to describe the ancient sites of the world individually and in isolation, it can be difficult to get any kind of overview of how they compare with each other. But there is an architectural pattern here.

To the northeast of Göbekli Tepe in the Caucasus Mountains, there are numerous stone dolmens with the exact same configuration. Several of these are also decorated in relief with what looks for all the world to be a megalithic post-and-lintel portal! It is tempting to attribute this coincidence to another arm of migration with a cultural memory going back to the "Motherland" of the Tigris & Euphrates area that contains Göbekli Tepe. There are reported to be dolmens of the same design as far as in India.

Schmidt was quite excited about the discovery at Göbekli Tepe of what he called porthole-stones. Usually rectangular, they are highly stylized. He wrote, "The stones also exist in monumental size which allow a person to crawl through the porthole." In the beginning of the Göbekli Tepe project, these objects were called 'portable pillar bases' because of similarity with the two pedestals in enclosure E. However, Schmidt reported that, ". . . during the 16 years of excavations many fragments of such stone have been discovered both in layer II and III, but no situation has ever been found confirming the suggestion that the foot of the pillars had been fixed to such portable stone frames."[22]

There is one particular such stone from Göbekli Tepe on display at the Şanlıurfa Museum of Archaeology. In an example of what sometimes happens, the old label stuck and it is described as a pillar base. After Dr. Schmidt passed away, no one thought to question that interpretation. This writer, having a different idea, was startled to find it displayed in the new museum in its current random position. Considered as a porthole stone, in the way Schmidt describes the transitional element of passage between interior and exterior, a rotation of this stone has significant archaeoacoustic implications. If one sees lips and a mouth represented here, there could hardly be a more expressive way of highlighting the place where sound comes out of the sanctuary. Thinking again of the isolation of Göbekli Tepe, one can easily imagine sound—pre-planned and engineered just like the stones were—projected from a portal like this one, thundering across the heads of gathered tribes. Or perhaps it was enough for a supplicant to get close and listen to the voices coming from the inside. There is not enough floor space in any of these monuments, Malta or Turkey, to comfortably accommodate at one time a gathering of all the members of a community that had enough population to build one—even if they were all standing up.

From a mechanical standpoint, what about the form and design of these places lends itself to the manipulation of sound?

4.4 Acoustification/Intention

The best and most necessary architectural feature for the purpose of good acoustics is a smooth hard surface. It's why we sound better when we sing in a tiled shower. Following a group of children into a concrete underpass quickly reveals what happens when they discover they can make echoes. Many of us have experienced an unpleasant audio overload in a crowded restaurant where they have taken no measures to damp down the racket with carpet or upholstery. Enclosure by hard reflective surfaces is what makes a space resonant. It could be concrete, tile, glass or even drywall. An empty uncarpeted house can produce wonderful echoes, particularly if it has a vaulted ceiling. Wood paneling and brick are not as good because they are usually more porous and tend to absorb the soundwaves to some extent.

The ancient builders had their hard stone interior walls, particularly since they plastered them, both at Göbekli Tepe and in the Malta temples. Where they didn't use bedrock or stone slabs on the floors, they tamped crushed stone into hard smooth pavement.

We have seen that the Hypogeum's "Oracle Chamber" has characteristics that apparently project sound energy in a highly focused manner. The builders in Malta even left us models of the temples. But can we say with certainty that the ancient architects came up with the design to intentionally boost the super-acoustics of the space? Of course we can't. What we can do is acknowledge that they kept using the same designs and techniques that continued to do such a thing.

These people of prehistory were clever and observant. Our predecessors were not strangers to natural caves. And they certainly would have noticed a difference in the sound of their voices as they passed from the outside world into enclosed spaces with hard reflective surfaces: so different, so odd. Having identified where the echoes were best, they could logically have made a connection with the physical features of the space like a naturally rounded dome or tray-stepped ceiling. There is another important feature which is well worth examining in more detail. It is the fact that they built with limestone.

4.5 Materials

Everywhere it is found, limestone is valued as a building material. The Parthenon was made with it. Egypt's great pyramid and the Sphinx were finished with it. The Holy city of Jerusalem was built in limestone, including King Solomon's Temple. The type in Malta that is called globigerina is extremely beautiful, taking on varied colors from the sunlight in different conditions. It is a stunning essential core of the ancient megalithic temples. There, the Phoenicians also built with it. Romans, Moors, Medieval Nobles and Crusading Knights all left their architectural mark on the islands with the Maltese limestone. With fine even texture and uniformity of color, the globigerina comes out of the ground moist and soft enough to take a mark impressed with a thumbnail. Then on exposure to sun and air it forms a hard protective crust. Once this crust is broken, however, disintegration ensues. The stone draws up water like a sponge. Over time, if not protected, it crumbles and dissolves away like a wet sugar cube.

By definition, limestone is a rock that contains at least 50% calcium carbonate in the form of calcite by weight. It is not volcanic. Limestone has what geologists call a biological origin. It is a sedimentary rock that usually forms in clear, warm marine waters from the accumulation of shell, coral, and other organic debris under the pressure of the sea above it. It's not unusual to find fossils in it and recognizable shells, even where it is now exposed high above sea level. In the ideology that we can call "indigenous", this stone could easily be associated with things that had life. We could understand a concept in the lore that surrounded its use, of relating it with the bones of the earth.

The geology of the Mediterranean/Levantine area hints at ancient basins and trenches that seem to have been vast depositories for the forming of this fine limestone. Malta is actually made of layered deposits of two kinds. A rough harder limestone which is called coralline is sometimes used on the outside walls of the monuments, while the softer globigerina, named for the planktonic creatures *Foraminifera Globigerina* from which it was formed, is normally reserved for more showy placement. It is possible to quarry fairly easily with primitive tools. The beautiful even texture makes it perfect for carving. And it is highly resonant.

It seems that the resonant properties of this stone were known in historic antiquity. It was the secret of an acoustic marvel around 2400 years ago when the rows of limestone seats at the theatre of Epidaurus reflected the voices of the actors on stage, reaching as far as the back rows of the theatre. The Romans selectively used limestone in the Colosseum. It was too heavy to use for the entire structure, so brick was used for the bulk of the building. Why would the Romans even bother with the limestone? To enhance the roar of the crowd. Think about what happens in a sports arena today and how the cheering drives excitement.

The painted caves of the Paleolithic are limestone. Göbekli Tepe is made of limestone, quarried very near the site of the monuments. Although the outer surfaces have been exposed over time to different elements, the author can attest that the inside of the building stone of Göbekli Tepe and the globigerina of the Maltese monuments looks exactly the same even under magnification. Karahantepe is built of it, its subterranean spaces carved down into it just as Hal Saflieni was. If enclosed with their tops on, the rooms in these sites will echo -– just like the painted caves, just like Malta's temples and just like the Hal Saflieni Hypogeum . . . because that's what this limestone does. Were the builders aware? Was that one of the secrets? Did the sonic characteristics of the stone, absorbed from ritual use of caves, touch the mind of some mystic diviner?

Göbekli Tepe and Karahantepe are located in the foothills of the Taurus Mountains: the largest and most important karst area in Turkey. Karstification is the dissolving of sedimentary stone by the action of water as it makes its way toward sea level or other lowest point. All the best caves with their stalactites and stalagmites are in limestone, formed of minerals leached from the dissolving stone and deposited drip by drip. They are also generally the easiest caves to walk into. The limestone of the Taurus range is heavily folded and thrust because of the collision of the European and African tectonic plates. It is riddled with thousands of caves. People might have been spending a lot of time in these caves as the last Ice Age ended, 12,000 years ago. Were they listening?

The Taurus Mountains, (so named for the bull whose bellowing was so impressive?) are a place of many ancient storm-god temples. Torrential thunderstorms in these mountains were deemed by the ancient Assyrians to be the work of the storm-god Adad to replenish the Tigris and Euphrates rivers.

It would be interesting to know whether limestone, by its chemical nature, has a natural "most responsive" resonance quality. The size and shape of the room, whether modern or ancient, will influence at what frequency it has its highest resonance.

It is enormously tempting to jump to a conclusion about this.

4.6 Incentive and Project Management

The logistical challenges of building a megalithic monument can be imagined. Problems would have been insurmountable without organizers and specialists. Many folks who have attended a homeowners' association board meeting can sympathize with what would happen without respected authoritative leadership.

"While we know the when, where and how of these buildings and who the people were who constructed them, we do not know why they were of such importance to their pattern of life" says Prof. England, speaking of the Maltese monuments. "There is little doubt that man at that time was already conscious of his temporality and as such the buildings he erected for his personal use utilized transitory materials, while when building for gods or sacred deities, there was a definitive collective effort to defy time and build for eternity."[23]

Lest we think it was an accident or intended as a one-off experiment, imagine what was going through the head of the first architect. There is much more involved here than an inspired idea of what it would look like when it was finished. Somebody had to think about organization of labor, project management, and continued motivation. Executing a monumental design like this requires clear communication skills and development of a complex language. It was necessary to think through every step of assembly, and pre-engineer every move for every great stone from quarry to installation.

The incentive to stay in place and build the first shrine had to be enormous. The whole enterprise would have called for a complete social reorganization. There surely must have been moments when the wisdom of such radical behavior was doubted, argued and disputed. The implication is that there must have been a highly regarded "voice of authority" to get it all started and keep it moving against any opposition, as well as an ultimate unquestionable "spiritual" rationale as motivation.

That brings us to a look at the nature of the sound that might have been made inside these ancient monuments.

Chapter 5

Music

As a species, homo sapiens has been making deliberate noise for a very long time. Just as birds and other animals alert each other, the need in early societies to keep the group members informed probably involved various vocal signals long before words and a spoken language developed. The shrill call of a whistle and the message of the "jungle drums" or even banging on lithophonic stones were good early warning detection systems in an untamed world. At some natural point, humans went beyond making sound related simply to survival.

Lawrence Christian commented astutely on a paper about this subject: "I have often wondered if some of these ancient sites were places for the dramatic expression of origin stories. That is, through visuals and sounds the next generation of a local culture was initiated into mysteries and stories of their origins. Today we might do it through film, or even in a multimedia Powerpoint. The technology changes, but the purpose remains the same—we are telling our stories as dramatically as we can."[24]

Megaliths, Music, and the Mind
Linda C. Eneix
Copyright © 2024 Jenny Stanford Publishing Pte. Ltd.
ISBN 978-981-5129-25-0 (Hardcover), 978-1-003-46823-3 (eBook)
www.jennystanford.com

Some difficult questions arise when this subject is introduced. Dr. Rupert Till, a Senior Lecturer in Music Technology, presents some of them. "Music is often defined as organized sound (and silence), but at what point does a sound making device become a musical instrument? When does something become music? How is this different to sound? If you clap your hands, are you applauding, dancing, musicking (music making) or doing all three?"[25]

Hearing is every bit as powerful in stirring emotion and as central to our understanding of our surroundings as is the sense of seeing. When did we lose "sight" of that? Curiously, the English language does not even have a proper verb for sharing sound. We can say "let me show you something" about the visual, but "hear you something" or "listen you something" just doesn't work. The closest we can come is "let me play you something", which implies that a device is involved. That is only appropriate when one is using such a device. The other possibility might be "let me sing you something", implying a personal performance from the speaker. There is a linguistic deficiency that says something about modern development.

We are hopelessly removed from the pre-technological world of our ancestors. Yet, even today, sound and music have the power to move us deeply. Opera is a fine example: huge drama on a stage with music that doesn't stop from beginning to end. Grammy-nominated Tenor Joseph Calleja puts it this way: "Opera deals with emotions that are too big to be spoken. They have to be sung."

While we consider an opera performance to be for the benefit of the audience, the singing is also a release for the performer who is doing the vocalizing. It may be more than just emotions being released over those vocal cords. There is a physical aspect that is rarely thought about these days.

A patient is waking from a major surgery. Nurses are moving her to a different bed and she cries out in pain. "Why are you yelling?" the nurse asks. "Does it make you feel any better?" The patient is a modern woman who has been taught that good girls do not make noise and disturb people. She interprets this remark as sarcasm. A resentment rises as she fights to stay quiet. In fact, the act of making the noise came automatically; it was a release; it *did* help. And that nurse may have been asking a legitimate question—because medical research is seriously opening up where the effects of sound are concerned. (Meanwhile, in most modern Western society, little boys are encouraged not to cry and teen girls dab daintily at their tears, while psychiatrists know that real release comes with the vocalized sobbing).

Music Anthropologist Prof. Iegor Reznikoff proposes Hal Saflieni as a surviving link between Paleolithic painted caves and Romanesque chapels . . . "That people sang laments or prayers for the dead in the Hypogeum is certain, for (a) it is a universal practice in all oral traditions we know, (b) at the same period, around 3,000 BC, we have the Sumerian or Egyptian inscriptions mentioning singing to the Invisible, particularly in relationship with death and Second Life, and finally (c) the resonance is so strong in the Hypogeum already when simply speaking, that one is forced to use it and singing becomes natural."[26] (See also Bonus Material in this volume.)

A live rock concert that sends the audience into a frenzy is an extreme, but relatable example. The principle is the same. We wouldn't be making sound and listening to music if it didn't do something for us. The equivalent in prehistory was very likely the ceremony of ritual and story-telling that explains the mysteries of human existence. If we consider that the basic dramas of life are birth and death, the character of sound and what was being done with it in ancient monumental spaces takes on new relevance.

5.1 Instruments

Out on the stone cliffs in the middle of nowhere in Malta, the wind howls in the crevices in eerie harmonics that rise and fall. It is hypnotic to listen to this primeval unstructured melody. There is a timeless lulling to it that makes even the most distant intrusion of modern noise discordant.

Natural movement of the air across a hollow tube of reed would have captivated the earliest thinkers. In the attempt to harness it, wind instruments were likely among the first noise-makers made by human hands. The documented practice of rituals involving the ancient Greek hoofed-god Pan in historic classical times provides a suggestive window into times even earlier.

Archaeologist Dr. Nektarios Peter Yioutsos has been studying the ceremonies performed in sacred caves in worship of Pan and his female companions, the Nymphs. They are described vividly in the Dyskolus of Menander. To begin, Yioutsos observes, "Evidence from a large number of caves proves that these breathtaking underground halls were the site of recurring ceremonies; and the large number of cave drawings, usually in extremely unreachable parts of the caves, illustrate man's efforts to encounter the supernatural deep under the ground. Archaeological research in Greece verifies the ritual use of caves already from the prehistoric period."[27]

Pan was a wild and lecherous sort of nature deity associated with flocks and shepherds, and nearly always depicted with his reed pipes. According to Yioutsos, he is connected to natural sounds, echoes and loud noise. We get our English word "panic" from his name.

"For the sake of Pan's love of noise a ritual performance to Pan and the Nymphs involved the production of various sounds," says Yioutsos. "Probably the aural qualities of caves responding to these sound signals wouldn't have passed unnoticed by their followers. To approach this god the ritual protocol demanded noise and the resulting resonating and echoing effects of caves would have been regarded as signs of his divine presence.

"Furthermore, when there was no recorded experience of the god's presence in a natural space, one could perform a sort of evocation ritual in order to provoke these auditory effects to happen by using various noise-makers, such as musical instruments, the sound of clapping hands and the sound of his voice. If the landscape responded to the artificially produced noise by sending sonorous sounds and echoes back, these aural features could probably have been regarded as indexes of Pan's strong or weak presence. Moreover, during a Pan ritual the participants through their noise-making, dancing, playing music, feasting and just having fun, they were forming a system of ritual action becoming themselves agents of god's epiphany.

"Pan had an intimate connection to echo and noise-making, and sound seems to have been an important factor in antiquity in the determination of caves as abodes of Pan and the Nymphs. These underground spaces may have been selected and even modified somehow in order to conduct and manipulate sound to produce various sensory effects during rituals. Recent archaeoacoustic techniques, applied in ancient sacred grottos and other constructions, have paved the way for new research insights in ancient cult and ritual practice, offering the potential to enrich interpretations of how ancient buildings or natural spaces were perceived as loci of divine presence and worship."

Excavations around the world have produced bone flutes and whistles that go back as early as 40,000 years ago. We'll never know how long people were whistling even before that. Scientists believe that birdsong came about as a way to defend a territory or to attract a mate. Maybe the same is true of people.

Current Archaeologists on the Göbekli Tepe project have been trying to interpret the possible meaning of several so-called "smooching" stones, recovered from the excavations. Could those pursed lips not equally be representing vocalization or even whistling? Unless we were looking for evidence that sound was part of the prehistoric scenario, would we know it if we saw it?

An estimated 10,000 of the modern world's people are still using a sophisticated whistling language to communicate across long distances. The whistling can be heard more clearly at distance than voice, even delivering clear messages between shepherds from mountain to mountain. Whole conversations and chats unfold without the use of any devices. Where is this ancient tradition found? It survives in Northeastern Turkey. There is a similar whistling communication used in the Canary Islands, although the signals in these "languages" are not the same.

Along with the wind instruments came percussion, with the human body as a sound maker, hand clapping and foot stamping. It is not known when the first actual drum was made with a stretched membrane. Traces of drums have been found in Neolithic period excavations.

Vocal cords were warbling long before there was a piano or a violin. Human voice was the first stringed musical instrument. Very early in human evolution, probably before we earned to speak, we learned to sing. Obviously we must have liked it because we have continued as a species to do it. Every indication is that making music has always been a universal human activity.

In the days before writing was invented, songs were the best way to remember and carry stories and histories. The person who knew and preserved the songs was the keeper of knowledge and of memory: the whole identity of the community. The songs were handed down from generation to generation and, in verse, were less likely to be corrupted along the way than mere story-telling.

A wonderful modern example is the Grand Song of the Dong ethnic group of China's Guizhou Province. The Dong people are described as a largely self-sufficient agrarian group with a population of less than 3 million. There's a deep-rooted belief among the members that "rice nourishes the body and songs nourish the soul."[28] Grand Songs are the vehicles of keeping knowledge from a time before the Dong had written language. All the historical stories and knowledge of the Dong group are handed down orally from generation to generation by means of songs that are taught by masters to choirs of disciples. Having been passed down through generations for over 2,500 years, the Grand Song is a cultural artifact so invaluable for humanity that it was designated by UNESCO as World Intangible Cultural Heritage in 2009.

Clearly, by the Bronze Age at least, music was an organized and important part of life. Evidence for the ancient lyre goes back as far as 4,500 years ago to ancient Mesopotamia in the area known today as Iraq. The Bible tells us that David played one for Saul.

Music and dance, celebration and ritual filled a purpose in their use. The practices did far more for them than just keep them entertained. They gave everyone a place in the world and a sense that they belonged.

According to researchers, a key function of music in early human groups was to strengthen bonds between group members. The idea is that these bonds led to more cooperative behavior.

It makes sense that each family clan or tribe would have its own "anthem". Each would have its own history and memory song. One could imagine that a gathering of clans at the world's oldest meeting places would have included a concert of clan songs. A song could easily have been a unifier, easing communication between tribes that didn't necessarily use the same words.

> **Exercise**
>
> Explore the group bonding of a shared sound-making experience. Suggest a trial rhythm session test to your group. Percussion does not require special equipment; hands on thighs and fingers on a table will work. Give no instructions. How long does it take for a particular rhythm to solidify and be followed by the group together? How many participants produce variations within the framework of the basic beat? Do the participants look at each other during the session?

As we have seen, monumental ritual spaces go back in time far earlier than the great pyramids and temples of Egypt. From the ancient British Isles down the Atlantic coastline, sweeping back in time across the Mediterranean to Spain, Sardinia, Malta, Cyprus—all the way back to the Mother of all monuments in Anatolia. It is very difficult to imagine that they could have been used in complete silence. What is the real evidence that music was part of the Neolithic scenario? Archaeologists would need to be very lucky to find anything like an instrument because the organic materials would be long gone. Anyway, if the Göbekli Tepe shrines were deliberately buried and everyone left, as the evidence indicates, then it makes sense that the instruments would have gone with them.

Art helps point to the answer. Excavations at the 8,000-year-old Çatalhöyük site, not so far from Göbekli Tepe, have produced fresco paintings that show people apparently dancing and holding round objects that are likely hoop drums (see Fig. 3.4).

Malta's Hal Saflieni Hypogeum, cut from the living limestone rock of that island, can help us understand the physics of vocalization in that space. According to Katya Stroud, "one of the earliest and longest standing myths is probably the idea that the Oracle Niche only reverberates and echoes when a male speaks or chants into it. This is not correct. Female voices at the correct pitch have also produced the same effect." Nevertheless, female voices are rare at the levels which have the most impact in the Hypogeum. The maximum resonant effect requires a baritone voice. A man can sing it far more easily than most women. Some females can get their voices down to the required level but it isn't comfortable for very long.

Earlier, we saw that so many of these very ancient places like Malta's Hypogeum, Newgrange and other passage tombs in Ireland and England all resonate at a common frequency within this "megalithic range" of 90–130 Hz. By very preliminary measurements, it seems that enclosure D at Göbekli Tepe fits the pattern as well. But what exactly does that mean? One hundred ten (110) Hz is the equivalent on the modern Western musical scale of the A2 note.

Prof. Reznikoff explains it thus: "Considering the strength of the reinforced sound, its duration, the number of possible echoes, we can conclude that the Hypogeum has 'good' acoustics for lower sounds. However, if the resonance of the main pitches A2 and E1 is very impressive, the resonance is still good on frequencies up to respectively the fifth E2 and the octave A3 of A2. The main range of male voices is amplified by the space. The weakness of the resonance, as compared to some parts of painted caves or to some Romanesque chapels, is that high frequencies, e.g., harmonic sounds or female high voices, are quickly dumped down, as expected because of the rather low ceiling of the rooms."

The person who does not have a musical background will appreciate more definition of these terms. We know that sound is vibration. Resonance is the quality in a sound of being deep, full, and reverberating. The Oxford dictionary defines frequency as the rate at which a vibration occurs that constitutes a wave, either in a material (as in sound waves), or in an electromagnetic field (as in radio waves and light), usually measured per second. As we know, a hertz or Hz is one cycle (or wave) per second.

A hide hoop drum can be played within this range. Drum and voice are sound sources that we know were around in very ancient times. During the 2014 testing in the Hypogeum, some other instruments were tried, such as a cow horn, a conch shell trumpet, and a very old Maltese folk drum that is played by the friction generated from rubbing a stick twisted into a hide that has been stretched and framed (bizarre but effective). These didn't have the same vibrancy because their frequency level didn't trigger much resonance in the space. They were not *tuned to the room.*

> **Exercise**
>
> See if you can source and play the A2 note with voice, instrument or phone. How many participants can vocally replicate that tone if they hear it?

5.2 Melody

It would be worth quite a lot to some people to know what ancient singers came up with for melody, although that's impossible. What we have to go on for clues would be the effectiveness of the low notes. It seems logical that in a venue like the Hypogeum, there would be a period of maximizing of a single note to bring it to full force, combined with a period of slight variation so that the returning echo of one note could be incorporated with the next. There may have been some mimicking of the natural sound of wind that one hears in the stone by the sea cliffs. The ancient Grand Song of the Dong people is also suggestive, sung a cappella in layers laid over a bass drone that fits the megalithic requirement.

Musicologist Dr. Stef Conner writes about reconstructing hypothetical Mesopotamian "proto-music" with a translation from cuneiform writings and a very detailed methodology involving something called "phylogenies" and musico-linguistic rules as a basis for song reconstruction. In her paper *The Score of Babylon*, Conner summarizes: "It is only by framing extant interpretations of cuneiform tablets, provided by musicologists and Assyriologists, within a methodological armature derived from comparative linguistics and musicology, sensitive to the facets of Babylonian poetics, that we can come close to understanding the materials out of which Mesopotamian music might be reconstructed. By contrast, a circumscribed intra-disciplinary examination of such materials would remain as fragmentary in its conclusions as the tablets themselves."[29]

Of course, we have no way to apply this methodology to prehistory. We uneducated listeners to Dr. Conner's performance sample of her work and the translated outcome, however, have found that ancient Babylonian lyrics are as expressive and emotional as the most weepy country ballad.

It is in Hungarian folk songs that Humanities Prof. Irén Lovász identifies the inherent, anthropological essential characteristics of singing itself. Dr. Lovász is also a professional singer known widely in Europe. In the text for the CD booklet of her album of folk songs for healing, she explains: "Through singing we experience different dimensions of time simultaneously, recalling as we do events and feelings from our own life, and at the same time participating in the eternal, timeless dimension of existence. This dance between past and present creates an archetypal, ever-present experience that lies at the heart of the healing power of folk song singing.

"We encounter physical, mental and spiritual benefits through singing all at the same time. The physical and physiological basis of such music therapy is the fact that our instrument is our own body. Through the act of singing, we create a flow of air that moves both inward and outward, activating our vocal cords and using our oral cavity as a resonant chamber. This process influences the vibrational and energetic levels of our entire body. Singing is, therefore, a manifestation of life force, akin to the very air we breathe. On an emotional and spiritual level, the creation and expression of art and poetry serve as a means of self-healing. The sung texts have connotations and levels of meaning that can evoke a strong mental and conscious effect. The essence of singing is to make the outsider inner and to project the inner world outwards. Great music involves a harmonic interplay between spiritual/emotional and physical vibrations."[30]

Once again, a "primitive" approach is bringing new understanding of how music and singing were regarded in antiquity, as well as what they continue to be able to do for us today. In this lies the recognition, recovery and preservation of something important that much of the modern world is in danger of losing.

As articulated by Dr. Lovász: "Physical sound waves resonate the strings of our soul. The energetic vibration of a healthy person is harmonic. When we get sick, however, this balance is disturbed, and we need to restore it in order to heal. Singing has a potent therapeutic role in this regard. With the help of our own vocal cords, we use our body as a resonant cavity, harmonizing our own physical and mental vibrations. By utilizing our own vocal cords and our body as a resonant chamber, we have the potential to properly harmonize our physical and mental vibrations." (See also Bonus Material in this volume.)

On a personal note, the author read the English translation of this CD text with a sense of awe. Without any musical training as a child and a failure to self-learn how to read it, something was triggered. Once, following a serious illness, there had been a moment in recovery when she realized that the music was coming back into her head. That was the first time she knew that there was normally always something floating around in there. It took the absence of it to bring the realization that it was a natural part of the healthy state.

Göbekli Tepe was also a site of intense religious experiences that reinforced beliefs and social networks. While under construction, even without roofing in place, just the sound of hitting those limestone pillars to carve them would have been like a rhythmic ritual! Did they perhaps have a sense that they were going to wake up the spirits of the underground?

For the workers who were preparing underground halls like the Hal Saflieni necropolis, the enclosed space would have created a spectacular sound environment. With antler picks and stone mallets, an estimated 2,000 tons of stone were removed to carve out the spaces. What the workers must have made of the thundering cacophony that had to be echoing all around them, we can only imagine.

It has been observed that the interior of some of the standing stones at Stonehenge were intentionally polished into a concave shape. Recently there has been quite a bit of research that hints at a ceremonial and acoustic purpose for this. The construction of the sarsen ring at Stonehenge ushers in the Bronze Age on the timeline of ancient building, as well as the advent of history. Acoustically, we can make an association with the sounding of a shofar in Jerusalem, the call of the muezzin in Mecca, bells tolling in the churches of Europe.

We need to examine human consciousness, not just in the intellectual state that we cultivate today, but also in the more mysterious states that, in some circumstances, become the essence of religion. Then we can understand if and how religious experience, belief and practice may have been the stimuli for revolutionary changes of the Neolithic period.

*The strange musical hum
emitted by Stonehenge
if a gale of wind is blowing
can never be forgotten.*

From *Tess of the d'Urbervilles*, Thomas Hardy, 1891

Chapter 6

The Mind

By this time, the religious connotations of the advent of monumental building are obvious. Although we have no documentation of what came before, we should have some understanding of the beliefs of ancient times in this part of the world—the times that are in the Bible. We know about Canaanites and Hittites and Sumerians.

There exist these ancient places in the world, like the temples in Malta and the buried shrines of Göbekli Tepe that were already lost and forgotten when the earliest recorded histories were being written. If they were ever referenced in the scrolls of the great libraries of the past, those are gone. To get to the roots of the archaeology of sound, and the sonic story of the world's first monuments, we do well to try to figure out what those Neolithic people were worshipping.

6.1 The Spiritual Element

Ritual and ceremony, of course, are supposed to imply some sort of religious activity. The use of special sound in that

---

Megaliths, Music, and the Mind
Linda C. Eneix
Copyright © 2024 Jenny Stanford Publishing Pte. Ltd.
ISBN 978-981-5129-25-0 (Hardcover), 978-1-003-46823-3 (eBook)
www.jennystanford.com

activity makes more sense when we consider how it would fit into belief systems.

In primitive days, much of the time, the source of a sound was clearly understood, such as the hoof beats of a migrating herd or the rustle of shrubbery that might be smaller game. An unusual sound with no visible source would invite some fantastic explanations.

Thunder likely gave rise to identifications in prehistory that were as wild as the ones our children know: Thor with his hammer, angels moving furniture or some heavenly bowling alley. Nearly always, there is a supernatural being involved. The list of ancient thunder gods is long, since nearly every culture had one. Native American legends include Thunderbirds, Thunder Brothers and Thunder Wives.

In their book *Inside the Neolithic Mind*, writers Prof. David Lewis-Williams and David Pierce ask the question: "Is it possible to have a religion that does not entail belief in the supernatural?"[31] As summarized online, "Lewis-Williams and Pearce discuss the archaeological evidence from both the Near East—including such sites as Nevalı Çori, Göbekli Tepe and Çatalhöyük—and Atlantic Europe, including the sites of Newgrange, Knowth and Bryn Celli Ddu. They argue that these monuments illustrate the influence of altered states of consciousness in constructing cosmological views of a tiered universe, in doing so drawing ethnographic parallels with shamanistic cultures in Siberia and Amazonia."[32]

Prof. England believes that in Malta's Hypogeum, the reverberating echoed tones would have been interpreted in prehistory as the supernatural voices of the gods, with prescient prophetic answers to queries made by mere mortals. "Was this oracle perhaps the early precursor of the oracles of ancient Greece? Here the building becomes the speaker and the listener's body the sounding board, integrating what is heard with what is seen, for while the eye reaches outwards, the ear receives inwards. One should not doubt that the reverberating spaces of the subterranean chambers of the Hypogeum, which to this day resonate in the depths of our sinews, are the result of an organized design programme. The whole atmosphere, both visual and aural must have provided awesome sensations to whoever was present when the oracle spoke, for man has a polar structure of perception. These people must surely have utilized sound as a channel to link the secular to the sacred. It is not too daring to assume that the Neolithic temple spaces of the Maltese islands were specifically designed to enrich the aural atmosphere of their interiors and raise the minds of the occupants and users to higher levels of consciousness."

People of the ancient world made associations in a different way than we do. Theirs was a world that was full of spirits and powers that hadn't been unexplained by science. Everything was experienced right in front of them. Whereas in the 21st century, we have become accustomed to wrappers, neat packages and tidy shops, the mysteries of nature were interconnected all around them in a way that was profoundly personal because it touched their lives so directly.

It seems natural to humankind to presume that there is some force or forces that have more control over the ill-understood than we do. An atheist has to acknowledge the power of a hurricane or a tornado even if he doesn't want to think about it. The Greek, Roman and even Sumerian accountings paint a wild picture for us of whole pantheons, living above, on and under the earth and sea.

Details of the way of their world were written down, defining the beginning of history in the Bronze Age around 5,000 years ago. In time, history tells us, the voices of the gods came to be interpreted for the masses by special go-betweens who had a direct connection. In classical literature, we learn about the tradition in ancient Greece and Rome of consulting an Oracle. In those days, people would go to a special shrine (usually a nice echoing cave) and ask questions of the spiritual leader, be it shaman or priestess, through whom a god would speak. Then there was the Sibyl, who lived in a cave, and doled out prophesy. One of these is said to have guided Aeneas through the underworld.

For thousands of years before the invention of writing, people were explaining the world as they knew it and expressing their beliefs. One way to start comprehending prehistoric culture in more depth is by considering how people were treating their dead.

It is not clear exactly what activity was happening as regards the use of Göbekli Tepe although we know the enclosures were not residential. Human bones have been excavated there, but the context was lost in the rubble made when the sites were "closed" later in the Neolithic period. Complex religious ideas are expressed here, and it is very clear that these were holy places to the people who created them. But it does not look like they were meant for funerals. Not too far away at the site of Çatalhöyük, archaeologists have found intentional burials.

The proto-city settlement of Çatalhöyük came into being about 2,000 years later than Göbekli Tepe, and continued in use up until about 2,000 years before the "Temple Culture" of Malta. Sandwiched between these two monumental articulations of the Neolithic, the large cluster of attached family dwellings serves as a good stepping stone. The transitional timeline between the hunter/gatherer lifestyle and the agricultural communities that would grow into cities is illustrated beautifully.

Excavations at Çatalhöyük tell a very clear story about a burial practice that would probably horrify most of us. Essentially, the deceased were kept under the beds of the living. The remains of the deceased were first excarnated. That means that the flesh was removed, in this case naturally, by leaving the body out for the vultures to pick clean. Our society might consider this grim recycling, but things were different for people of the Neolithic. For them, nature was nature and everything was part of the whole cyclical system. One can reasonably assume that there was a high level of ritual attached to the whole process. The scavenger birds that span the artistic expression of so many cultures including at Göbekli Tepe are a solid part of this concept of returning the dead to nature.

The recovered bones were then interred in a shallow grave, typically under a sleeping platform built into the living quarters. When the grave was needed again, it was opened, the bones of the most recently deposited deceased were pushed aside to make room for the next occupant. With whatever ceremony or custom, the new interment was made and the tomb was resealed. Some of the excavated skulls had been plastered and painted with red ochre, perhaps signifying the color of blood and life.

In the heyday of Malta's "Temple Culture", disposition of the dead was in specially prepared underground mortuary shrines like the Hal Saflieni Hypogeum. As at Çatalhöyük, the bodies were de-fleshed elsewhere, and the bones brought to the underground temples for ritual treatment in large communal graves. These were side caves, each of which was then sealed by a vertical stone that could be removed when the time came for the next use. Archaeologist David Trump makes an interesting point about social implications of this variation from the habits at Çatalhöyük. "The dead were no longer being cared for within the family group but were absorbed back into the larger community in the vast underground cemetery which that community, by mobilizing the efforts of its members, had hacked out of the rock to receive them."[33]

The careful recreation of the symmetrical monumental architecture of Hal Saflieni signals that ceremony by the living was part of the process.

In precious undisturbed burial deposits, red ochre was found to be dusted liberally directly onto the bones. When Hal Saflieni was discovered in 1902 and "emptied" a few years later, there was not yet an organized science of archaeological excavation. Much information about context was lost when ancient bones of an estimated 20,000 people were brought out by the wheelbarrow-full and spread on the fields for fertilizer. One intact skeleton was found, but it is not clear if it was part of the Neolithic deposits or the remains of someone who found his way in later in time and never managed to get out.

Another mortuary site on the sister island of Gozo, belonging to the same culture, has given much clearer evidence as the result of modern excavations. It appears that in the Maltese islands there was some sorting of the body parts, long bones in one pit and kneecaps in another. Sometimes the skulls were kept piled up together. Children had their own place separated from the adults. With the great communal burials, one gets the idea of the bones being "planted" like seeds for some sort of cosmic renewal and rebirth. A single burial was found that had received special treatment. It was the intact skeleton of an arthritic female who was interred wearing her shell headdress.

In Sardinia, the many "Domus de Janas" (translation: House of the Fairies) tombs dating to between 5400 and 4700 years ago show the same custom of painting the walls and the deceased with red. Red ochre shows up everywhere in human burials, not only of the Neolithic, but as far back as a reported 90,000 years in a cave at Qafzeh, Israel.

In Malta, where ancient remains lie under practically everything, a small previously unknown prehistoric tomb was accidentally opened by road workers. It had been flooded, the water mixing with the red ochre of the burial. It looked so much like blood when it came gushing out that the workers called the police, thinking that they had found evidence of a recent murder.

Red is supposedly the first color that was perceived by the evolving human species. Five thousand years ago in the megalithic temples of Malta, even the plastered stone interior walls were painted red. Probably because of its obvious association with blood, prehistoric people likely considered red to be endowed with special powers. In fact, the word "magic" is related to the Anglo-Saxon word for red ochre. Babies were born bloody. People (and animals) who bled too much died. One of the first associations made would be that red was life and red was "special".

People of the Neolithic Age were living daily with life and death very much in front of them. The cyclical changes in nature: of regeneration and growth and bearing or fruiting, were the reference points for everything. Obviously, the earth was producing food for people to eat as well as the plants that fed the animals. Obviously, it was the female of each species that was giving birth and feeding her young. The association with something that we call "Mother Nature" would seem logical. If we want to, we can just about recognize it in the sculptures that have been found in temple and tomb and home across Neolithic cultures.

Some artifacts are obviously female and others not so much. As cult objects, some of them may be representing something more like abundance without gender. In fact, in both Malta and at Çatalhöyük, all the signs point to egalitarian society. There seems generally to be no differential burial practice for men or for women, other than that single important lady of Gozo.

Ancestor worship could have been part of their belief system. Perhaps there was a Great Spirit as in the Native

American tradition. We don't know. Were they worshipping Gods or Goddesses? For the purpose of this study, it doesn't really matter... except maybe for the tone of voice.

As an historical reference, we can turn back to those ancient traditions of Oracles and Sibyls, consulted as go-betweens through whom the gods spoke. Most of the original Oracles were devoted to an Earth Mother Goddess in Greece, Egypt and other parts of the ancient world. The ancient religion that produced the Golden Calf of Exodus fame was based on a trinity that changed from city to city, but always included at least one god and one goddess. Coincidentally, this is the religion that is frequently referenced with disdain in Biblical texts, with instructions to "break down their altars, smash their sacred stones." It is not difficult to suspect a remnant faith from far earlier times in the region of Göbekli Tepe in Anatolia that was still functioning for the Canaanites.

What might be more important is the conception of the earth as life-giver and thus sacred. Imagine that before there was a perception of "God in Heaven Above", and before the creation of pantheons that split the fateful territories of human existence between different entities, the Great Creator's realm was beneath the land. So intrinsic was the role of the earth to the way of life of the people of the time, that otherworldly sounds seeming to emanate from the stone would have had tremendous significance. The concept of a spiritual world within the rock is demonstrated in different parts of the world in rock art where painted figures seem to come out from cracks in the surface.

6.2 Magic?

Most university physics laboratories have a Kundt's Tube device for demonstrating the speed of sound. Within the tube, it is possible to make sound visible in the suspension of fine powder or dust. There are many video demonstrations of the phenomenon on the internet. It is fascinating to watch the molecules come to life and take form, shaped by the vibration of the sound generated in the tube. (By now the reader might guess where this is going.)

In his book *Stone Age Soundtracks*, Paul Devereux writes that to test a theory, Physicist Bob Jahn's Princeton laboratory made a working model of the Newgrange Passage Tomb. "Ironically, the passage is the megalithic equivalent of a Kundt's Tube." They used a special sort of dust, designed for the purpose, to stand in for the smoke or steam of a hypothesized ancient ritual situation. Devereux reports results, "Sure enough, when the experimental apparatus was switched on, the airborne dust particles clearly vibrated to reveal the acoustic standing wave within the tube when an intense beam of light simulating the sunbeam was shown through them, presenting a scaled-down version of what actually could have happened in the Newgrange passage."[34]

In this last remark, Devereux is talking about the Winter Solstice sunrise that enters Newgrange through the window box above the entrance.

It does not take much imagination to guess how such a thing would have been experienced in prehistory. Visible waves would have to be construed as something mystical. How easily sound might have been understood as the voices of spirits, ancestors or other forces! It would have contributed to the specialness of the place as these effects could not have been reproduced anywhere else, for example, in the open air or inside huts where people lived. It would be magical! Add some beer and maybe some medicinal herbs to the equation and . . . well, one gets the picture. And the person who knows how to produce such magic would wield a lot of power.

Strong sunlight entering a dark space through a small opening is required to make it work.

Figure 6.1 Mnajdra South Temple, Malta—dawn on Vernal Equinox, just before sun appears.

Coincidentally, the original entrance to Malta's Hal Saflieni Hypogeum, now enclosed by the visitor center, was aligned with the Summer Solstice sunset, allowing the last rays to penetrate all the way into the middle level, along a pathway washed with red ochre, flooding the red and black-painted central hall chamber and finally entering the finely framed back niche of the "Holy of Holies".

Above ground at Malta's Mnajdra Temple site, there is a full-blown calendar in stone that still functions after 5,000 years. On each Equinox dawn when there is no cloud, sunrise enters the temple through the main portal, floods directly down the main corridor and lights up the stones in spectacular fashion. Imagine the ceremony that could have been made of that. It was well worth getting up for even with open sky overhead instead of the enclosed space the structure must once have been. Happily, the daybreak alignments have not been blocked by the installation of protective covers over the site.

The Mnajdra calendar was not accidental. The interior stones are precisely placed to channel sunrise on both Solstices as well as the Equinoxes and everything in between. It's too perfect a system to have been random. Archaeoastronomers have found alignments of one kind or another nearly everywhere they looked, but even the archaeologists have accepted the veracity of this issue.

In Neolithic times, the importance of knowing where one was in the turning of the seasons was about more than knowing when to plant. An agricultural community that was going to thrive had to store surplus. Someone had to know how to calculate the number of mouths to feed and how long the available food resources were likely to last until they could be replenished—whether that was by new crops or new lambs. With no supermarket or convenience store down the road, awareness of seasonal change was a very big deal.

The regular patterns of the sun, the moon and constellations made up a fabric of predictability that provided a sense of normal in an environment that was in constant change with weather and life's challenges. The drama of pivotal points in the passage of time surely must also have been fraught with this connection between music and deep emotion. What rites were attached to the special solar events 5,000 years ago? Could the ancient people have been led to believe that the voices of their gods or something of the spirits danced inside here in the first rays of the sun?

6.3 Art

For most people ancient Egypt is where archaeology and antiquity start. The strange rituals of mummification and beliefs about Second Life captivate us. We know a lot about the ancient Egyptians because the climate in the region stalled decay and because their hieroglyphs documented everything in a written language that has been deciphered. The story of prehistoric people has to be pieced together from what has survived of what they left behind. Their art is expression that we can try to read.

Can the sculptures, carvings and paintings made in ancient times give us clues to the expression of sound? We have developed modern icons such as stacked horizontal lines that curl at the end to represent wind. How does a prehistoric mind that has not been trained in today's acceptable techniques illustrate the invisible?

The Neolithic: that mysterious time between the last Ice Age and the invention of writing was a time of spectacular human development. The people who lived in the "fertile crescent" in those days were smart risk-takers. And, as we have seen, they loved their red. In western Turkey, in the excavation of a pottery workshop that goes back 8,000 years, archaeologists found red ochre everywhere. It is on the tools and painted on vessels. There were bowls of unground chunks of the source mineral hematite that weren't used yet: prehistoric artwork that never happened.

Traces of the pigment that once colored the walls in Malta's underground Hypogeum 5,000 years ago can still be seen. One can imagine that standing in the central hall by the light of one or two of those moss and animal fat lamps could have been likened to being inside a living organ: perhaps the womb of an earth deity.

The mineral deposits for red ochre do not exist in Malta. The substance was so important to the people of the time that they were willing to leave the island and come back again to keep supplied, unless there was some sort of trade going on already in those days.

The most spectacular use of red pigment that remains visible from the Neolithic is the painted patterning on the ceilings of Malta's Hal Saflieni Hypogeum. These mystical designs were applied by artists 2,000 years before the birth of Nefertiti or Tutankhamun.

The chamber beneath the spiral/disc motif is the Oracle Room. The pigments are faded now and difficult to fully assess in the low light that is required for conservation. In drawings of the space published in 1971, however, it is clear to see that the patterns begin at a point coinciding with an opening in the wall.[35] This is the "Oracle Niche", shown also at the far right of the top photo in Fig. 2.6, that has long been associated with the odd acoustics of the Hypogeum. As earlier described, the room is shaped like a long gourd with no second exit. All sound travels back out into the more finely carved "public" spaces. The ceiling patterns get smaller in scale away from the niche, becoming denser, multiplying and flaring out at the open end of the room on the left. While there are other paintings in the Hypogeum, this is the only room with these connected swirls.

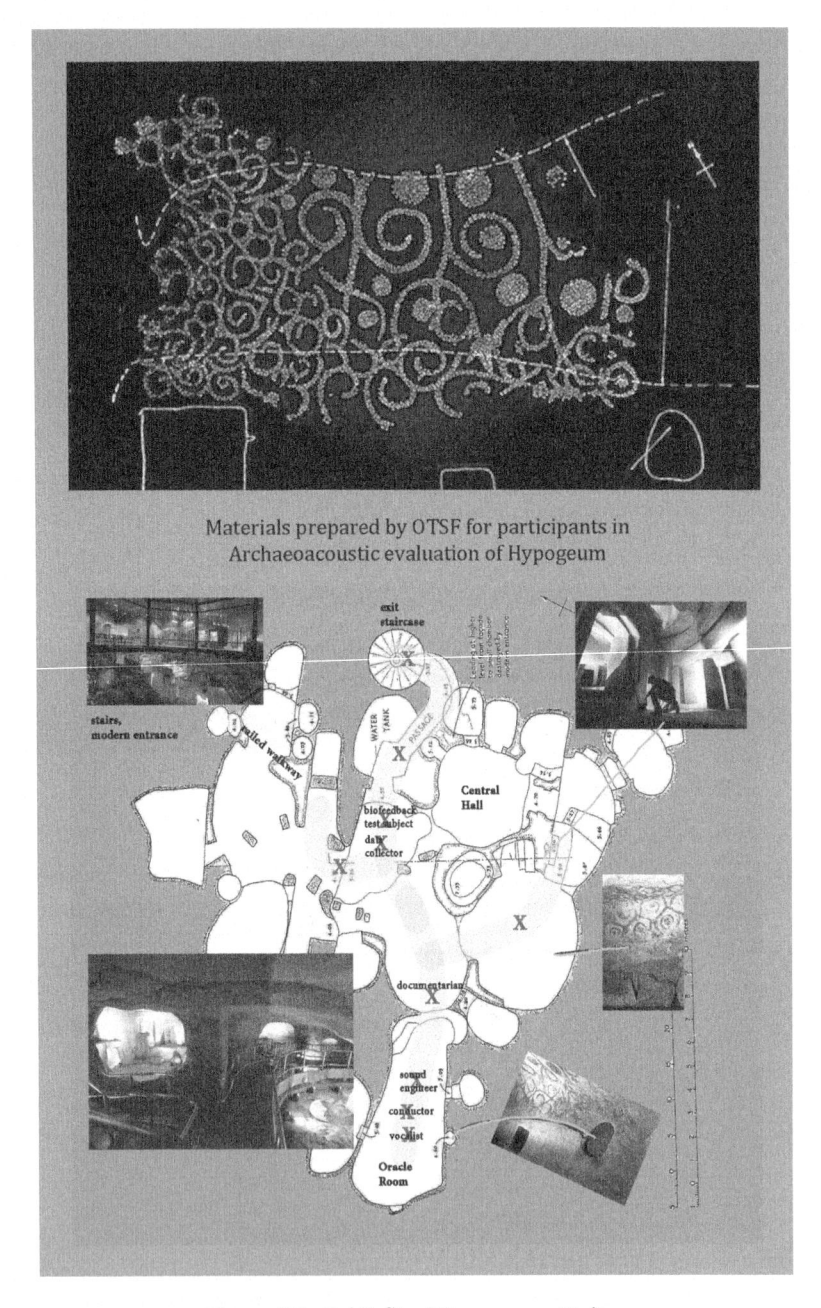

Figure 6.2 Ħal Saflieni Hypogeum, Malta.

Prof. Reznikoff considers them, "in relationship with the megalithic spirals in Western Europe (Brittany, Wales) that inspired the Irish Christian sacred art (crosses, sacred books) where spirals and their interlaces represent the Spirit that comes from the cross or the Heart of Christ and blows everywhere."[36] He suggests that the chamber ceiling may represent "the arborescent movement, the red discs being as incarnated fruits, of the becoming souls and spirits in the visible and invisible worlds."

Reznikoff is highly respected for his work in associating the painted images of Paleolithic caves with their location in the most acoustically vibrant parts of the caverns. After a personal evaluation in 2014, he shared his conviction about Hal Saflieni:

"The Hypogeum, caves, temples or churches, share the same meaning in addressing the Invisible by chanting, painting, possibly dancing, and praying . . . . The deep answer of the resonance changes the personal timbre of the voice and seems to be the sound of Mother Earth or coming from the Other World: worship introduces to deep levels of consciousness."

It appears that in addition to the color red, those Neolithic minds in Malta were also obsessed with the spiral. Exquisitely carved into the fabric of their monuments, the spirals appear over and over again, in a dozen variations. They join the carved stones of Ireland's passage tombs like Newgrange and Knowth, Piodáo in Portugal, Pierowall and Isle of Eday in the Orkneys as a "Neolithic signature". These spirals and markings are more than just decoration. They are saying something, expressing meaning. Can we imagine why ancient people would have gone to the enormous concentrated effort that is required to carve something in three dimensions like these large stone screens? Those who think it looks easy should try tracing the pattern with paper and pencil.

Figure 6.3 (*continued*)

Figures 6.3 "Screens" from Tarxien Temple, Malta once faced each other across a corridor.

136 | *The Mind*

"Sleeping Lady" from the Hal Saflieni Hypogeum. Note the seams in the skirt that depict the joining of two widths of loomed fabric. Also, the hairline seems to have been shaved back at the forehead in an interesting style. A very slight indentation on the brow hints at a decorative headband. Cane furniture seems to be be a well-developed part of the prehistoric scenario, although no physical trace has survived 5,000 years.

Figure 6.4 Malta's most iconic artifact of the "Temple Period".

Rock Art specialist Dr. Fernando Coimbra wonders if some of these designs can be the result of phosphenes, or other mental imagery or illusions caused by sound effects. He suggests that the prehistoric art that exists in some of these chambers could be related with eventual bodily experiences with acoustic origins. Considering his own extraordinary experience of feeling sound in the Hypogeum, Coimbra is led to wonder, "Did Neolithic human beings feel similar sensations, caused by sound, while staying inside chambered monuments?"[37]

An iconic sculpture that was discovered in the Hypogeum is affectionately called "The Sleeping Lady". There are other similar pieces in the museum collection with figures placed in sleeping position on beds, but none as intact or so exquisitely modeled as this little five-inch-wide masterpiece. Is it a votive perhaps?

Coimbra and others think that the prone figures could be suggesting the practice of dream incubation. "In fact, being laid on a couch could be related to the act of comfortably hearing reverberating sounds inside Hal Saflieni and getting any kind of bodily sensation or inspiration through them."

How interesting that in all the human representations from Malta's Neolithic Temple Culture, and there are more than 20 of them, while other features were modeled beautifully, not a single one of them has any ears showing. People who could sculpt like this could have made ears if they wanted to. One could almost get the impression that hearing was so important to these people that the organs for it were considered private—even to the point of carving the elaborate hairstyles of the day, or caps or hoods to cover them.

In contrast, the figure called "Urfa man", discovered not far from the Göbekli Tepe site in Turkey, has got good ears on him, as well as other anatomical features. But he curiously has no mouth. It isn't broken. It was never there.

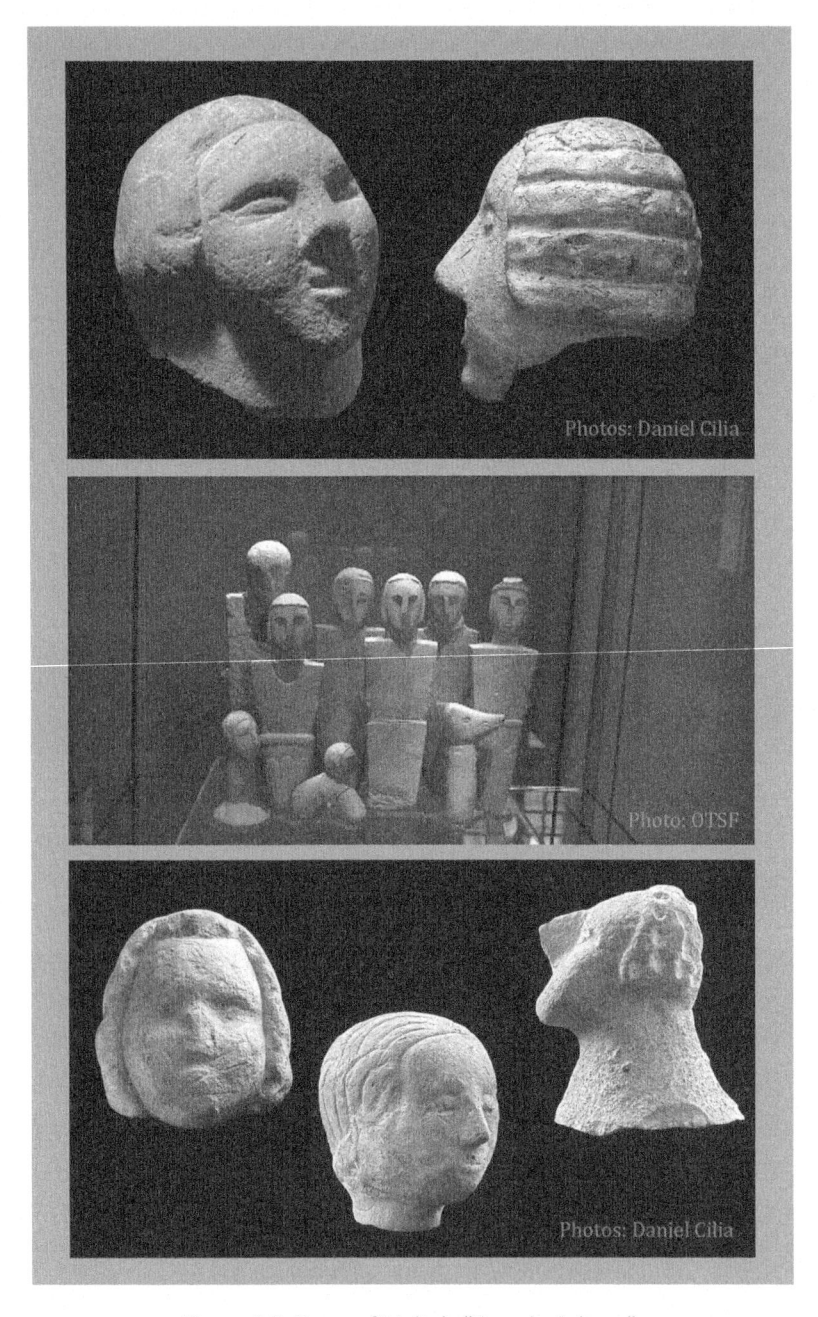

Figure 6.5 Faces of Malta's "Temple Culture".

Art | 139

Figure 6.6

140 | *The Mind*

Figure 6.7 Clockwise from left: "Urfa man", "column base", a "smooching head", indicia at throats of central anthropomorphic columns, Enclosure "D", Göbekli Tepe.

A similar head was found in one of the sanctuary circles, also conspicuously without a mouth. Its findspot suggests that it was placed as an offering during the "closing" of the site in prehistoric times, perhaps as if to say "We won't tell."

Much of the iconography of the cultic community of Göbekli Tepe is found in excavations spread throughout Upper Mesopotamia in representative settlements of the time period. It shows that the people had a complicated mythology, including a sophisticated capacity for abstraction. Motifs and symbols were inscribed into the pillars as well as on portable plaquettes found at many other sites in the region. Animals like snakes and scorpions, quadrupeds and birds are stylized. Pictographs of a sexual nature leave little to the imagination while several complex totem-like columns of stacked-up beings are very difficult to decipher.

The central anthropomorphic pillars of Göbekli Tepe's excavated shrines are marvelous expressions. Clearly suggesting human figures, the arms and hands at their sides wrap around the edges of each pillar to rest fingers upon the belly in front in the most formal of poses. Standing at the base of one is quite humbling. Both pillars of enclosure D have been preserved more or less intact and each reaches a breathtaking 18 feet or so high. The projecting portions of the T-shaped stones are interpreted as representing heads that loom over the stylized bodies below. Dr. Schmidt describes the uncovering of these magnificent pieces: "It was no surprise that hands and fingers became visible soon on both pillars, but also an unexpected discovery could be made: both pillars wear belts just below the hands, depicted in flat relief. A belt buckle is visible in both cases . . . From both belts a loincloth is hanging down covering the genital region. It is apparently made of animal skin, as the hind legs and the tail are visible." Since several clay figurines from another pre-pottery Neolithic site are wearing belts and all are male, Schmidt thinks it probable that the two statues in enclosure D represent males, as do so many of the carved beasts there.

Figure 6.8 In situ, Enclosure "D".

Art | 143

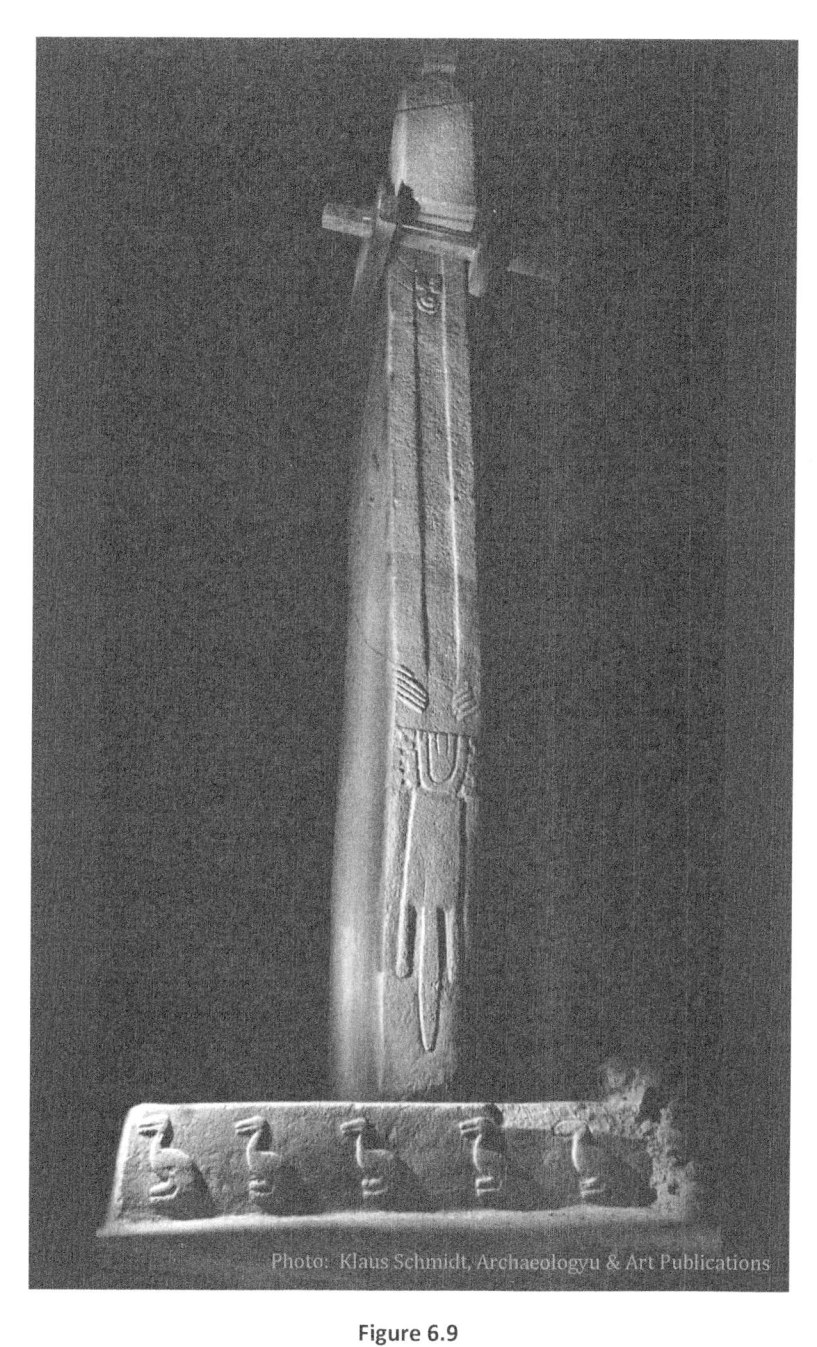

Figure 6.9

Figure 6.10 (*continued*)

Art | 145

Figure 6.10 Illustration created from full-size reconstruction.

Further evidence of a social system is hinted at in the artifacts. Spacer beads with one or more drill holes are reported by Schmidt to be common at Göbekli Tepe and at two other sites, but absent in others. "It seems," he says when examining the distribution, "that a special type of ornament was used by a specific group of people, or, in other words, had the role of being a visible sign for people involved in specific functions."

If these paired pillars are meant to represent twin gods, they are not identical. Perhaps their most significant variation is the features that are depicted at their throats.

While there are no facial features on any of the anthropomorphic pillars of Göbekli Tepe, on these two there are representations of necklaces or collars worn right where a voice box would be. Whatever they are, they were important enough to be carefully sculpted in the stone. More than just decoration, these designs are perhaps something like emblems of office or other distinct differentiation between the two figures. Could this possibly have something to do with voice and sound? Dr. Schmidt noticed that one of these carved columns "sang" when struck with a fist.[38] When tested on site in the presence of this author, the resonant nature of the stone was audible without the need for instruments.

Dr. Schmidt describes other humanistic sculpture like the previously mentioned "smooching" heads he uncovered on site. "These more or less naturalistically depicted statues seem to represent members of our world, powerful and important, but inferior to the T-shapes, who remain in mysterious, faceless anonymity. The T-shapes seem to belong to the otherworld; the non-stylized statues have the role of guardians of the sacred sphere." He believed that the art and architecture of Göbekli Tepe demonstrate that, "in Upper Mesopotamia within the society of Early Holocene hunter-gatherers, a powerful elite is observable, who is not visible in the former Palaeolithic societies." How did such an elite class come into being? We might wonder.

6.4 Neuroscience

Although missing the component of the ancient world, *Music and the Mind* is a very up-and-coming combination in neurological research these days.

A program launched by soprano Renee Fleming in collaboration with the U.S. National Institutes of Health (NIH), with the participation of the National Endowment for the Arts (NEA) provided inspiration for a deeper look into what has been happening with sound in neuroscience.[39] New information is flooding in every day about the complex relationship of music/sound and brain activity. According to Harvard Medical School neurologist and psychiatrist David Silbersweig, music activates many different parts of the brain. "The process by which we're able to perceive a series of sounds as music is incredibly complex." Says Silbersweig. "It starts with sound waves entering the ear, striking the eardrum, and causing vibrations that are converted into electric signals. These signals travel by sensory nerves to the brainstem, the brain's message relay station for auditory information. Then they disperse to activate auditory (hearing) cortices and many other parts of the brain."[40]

The revelations are stunning but none so exciting as one from Johns Hopkins University, where researchers have identified a relationship between music and dopamine release. There is a lot to say about how this can be applied to what was going on in prehistory with the raising of the first megalithic monuments and the sound behavior inside them. And there is obviously much more to be said when one begins processing these connections.

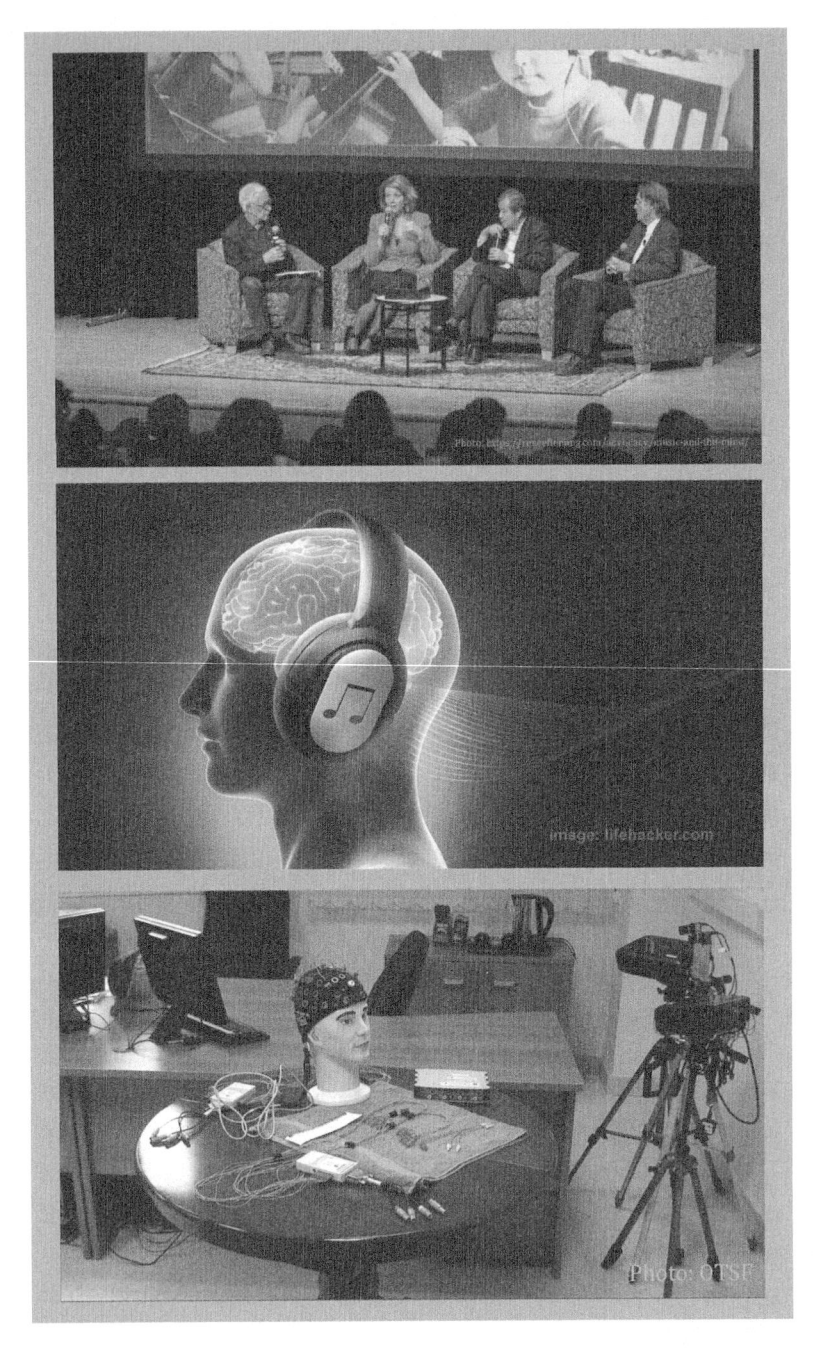

Figure 6.11

"Transcendental resonance" is the term for an altered state of consciousness triggered by the right sound. Buddhist chant is an example of a technique that has survived into modern times. Australian associations with "dreamtime" accompany the original ceremonial playing of the digeridoo. As introduced by journalist Cody Delistraty, "Through its haunting, yet powerful, sounds, the digeridoo evokes reverence and oneness with nature. The deep drone of the digeridoo allows us to relax and to reconnect with mother earth."[41] Were the ancient bagpipes of Malta and Scotland contrived with a drone in the same way?

Archaeologist Fernando Coimbra is trained in meditation and was able to describe what he experienced in Hal Saflieni during the 2014 archaeoacoustic evaluations: "Standing in front of the prehistoric paintings in room 20, the sound of a horn played in the Oracle Room was felt crossing my body at high speed, leaving a sensation of relax. The experience was repeated after some minutes and the result was similar but even more relaxing, accompanied by the illusion that the sound was reflected from my body to the walls."

Looking more deeply into his own experience, Coimbra reports, "Studies assessing brain activity by functional magnetic resonance imaging, in experienced practitioners of meditation, revealed that this practice 'increased activity in the prefrontal cortex, showing a large cluster with the point of maximal activation in the straight gyrus, covering a large part of the right orbiofrontal cortex as well as medial prefrontal areas'." The authors cited by Dr. Coimbra argue that this activity is related to the relaxed focus of attention, which allows spontaneous thoughts, images, sensations, memories and emotions.[42] Would such an experience be equally effective for the Neolithic populations? Why wouldn't it?

The part of the brain that deals with sound and music perception connects to other things the brain does: such as moving, planning, remembering, imagining, and feeling.

Dr. Robert J. Zatorre of the Montreal Neurological Institute is a pioneer in the area of music and the human brain. Could there be a field of study any further from archaeology? And yet, the findings of Dr. Zatorre, his team and others in the field present implications for this archaeoacoustic study that are staggering.

To begin: Zatorre's team confirms via PET (positron emission tomography) scan and fMRI (functional magnetic resonance imaging) that the brain can respond to the abstract stimulation of music with feelings of euphoria and craving. They cite direct evidence for the endogenous release of dopamine at the peak of arousal during music listening. Further, the report states that, "Importantly, the anticipation of an abstract reward can result in dopamine release within an anatomical pathway distinct from that associated with the peak pleasure itself. These findings help to explain why music has always been of such high value across all human societies."[43]

"Dopamine," as described by WebMD, "helps us strive, focus, and find things interesting. It affects many parts of behavior and physical functions such as learning, motivation, heart rate, blood vessel function, kidney function, lactation, sleep, mood, attention, control of nausea and vomiting, pain processing, movement."[44]

According to Harvard Medical School, Dopamine is most notably involved in helping us feel pleasure as part of the brain's reward system. Sex, shopping, smelling cookies baking in the oven—all these things can trigger dopamine release, or a "dopamine rush." This feel-good neurotransmitter is also involved in reinforcement. That's why, once we try one of those cookies, we might come back for another one (or two, or three). The darker side of dopamine is the intense feeling of reward people feel when they take drugs, such as heroin or cocaine, which can lead to addiction."[45] The overuse of drugs can cause the body to be less able to produce dopamine naturally, which leads to compound challenges for addiction recovery; a factor that is not widely known outside treatment circles.

Another report broadened the examination of physical response. "We used positron emission tomography to study neural mechanisms underlying intensely pleasant emotional responses to music. Cerebral blood flow changes were measured in response to subject-selected music that elicited the highly pleasurable experience of 'shivers-down-the-spine' or 'chills'. Subjective reports of chills were accompanied by changes in heart rate, electromyogram, and respiration. As intensity of these chills increased, cerebral blood flow increases and decreases were observed in brain regions thought to be involved in reward/motivation, emotion, and arousal, including ventral striatum, midbrain, amygdala, orbitofrontal cortex, and ventral medial prefrontal cortex. These brain structures are known to be active in response to other euphoria-inducing stimuli, such as food, sex, and drugs of abuse. This finding links music with biologically relevant, survival-related stimuli via their common recruitment of brain circuitry involved in pleasure and reward."[46]

These last paragraphs are worth reading again, in association with the focus of this study, because if it happens now, it was also happening in the Stone Age. (Note the phrase in this report: "subject-selected music". As we all know, not all music is pleasant to all listeners.)

Dr. Akhil Anand is an addiction psychiatrist who knows quite a bit about this startling suggestion. "Thinking of addiction as genetic begins with understanding that addiction is a chronic relapsing brain disorder. In many ways, it's no different than having a family history with heart disease or diabetes," says Dr. Anand. Research shows that genetics have somewhere between a 40% and 60% influence on addiction. Are there addiction genes? Dr. Anand claims that the genetic connection to addiction comes through inherited levels of dopamine. "Now, that doesn't mean that if you have the genes, or if you have family members that have struggled with addiction, that you're going to develop an addiction," explains Dr. Anand. "It just means you're more prone to it."[47]

Let's face it: addiction is a growing major problem in our times. The triggers for it are far more dangerous these days, physically, emotionally, and socially. Considering the impact of the previously mentioned migrations, an estimated third of the world's present population may have inherited a genetic propensity for addiction trouble. Exploring this matter in depth is beyond the scope of this volume, but the possibility must be mentioned. Such a concept is certainly provocative.

The basic human physiology of this neurological response to music has been the same for thousands of years. The crux of it is beautifully explained by reporter Cody Delistraty: "Interestingly, this dopamine activity occurs mostly in the caudate, which is a part of the corpus striatum, the part of the brain that controls motor skills (and mediates the reward center for important stimuli like water or food). This is why, at the climax of songs, when dopamine is released, we get physical 'chills'."[48]

The lab work cited earlier in this book in which volunteers listened to electronic tones (not music) in the "megalithic range" of 70 to 130 Hz. was very preliminary. And yet in three independent studies of stimulation by sound at the level that has been seen to produce the highest effect in ancient stone chambers, the changes in brain activity detected by EEG correlate to the cited report above. In other words, it is not just imagination or fancy. Whether today or 5,000 years ago, people standing in those spaces and hearing a low voice or a drum tuned within that range could experience a bit of a dopamine rush.

And that's not all.

They were as likely as not to have had some sort of spiritual experience.

Researchers at the University of Pennsylvania used brain imaging studies of Tibetan Buddhist meditators to provide empirical evidence for the specific mechanisms involved in a religious experience: a state which they have called Absolute Unitary Being (AUB). "AUB refers to the rare state in which there is a complete loss of the sense of self, loss of the sense of space and time, and everything becomes an infinite, undifferentiated oneness."[49]

This state of AUB, or being one with the universe is a neurological event, wired into the brain. It is a spiritual feeling that is common to all people. If you speak about your own experience with such a thing, others can understand what you are talking about. The parts of the brain involved are these same neural circuits that are processing emotions, as well as language and music.

When the experience is unfolding in a space that has already been designated as "sacred", the ritual & ceremonial environment of a tomb or shrine for instance, or a temple or a church or a mosque, it is safe to say that people would interpret their neurologically generated mystical states as contact with the supernatural, or the presence of God.

In prehistory, people who were standing in those resonant spaces and hearing a low voice or a drum tuned within that stimulating "Megalithic Range" could have associated the specialness of the place with a human desire to be inside it, just as the original architects did.

There is more.

Melody is the relationship of the notes to each other—going up or down in pitch. We recognize melodies, whatever key they are in, and we don't even think about it. People from different cultures who do not speak the same language, can recognize the same melody. If we heard the most remote

Amazonian tribe singing "Happy Birthday" in their own language, we wouldn't need to understand the words to know what they were singing. Researchers are exploring music as a way of communicating with creatures like dolphins and whales who do not even use language.

This ability to hear relative pitch in melody involves the right auditory cortex of the human brain. Zatorre's research has produced a startling parallel: "fMRI also shows activation of intraparietal sulcus, right parietal lobe, far outside the auditory cortex. This region is known to be involved in visuospatial processing and in visually guided special tasks, mental rotation."

As defined by Challenging Our Minds Weblog, "Visuospatial skills allow us to visually perceive objects and the spatial relationships among objects. These are the skills that enable us to recognize a square, triangle, cube or pyramid."[50]

That the two perception activities are so closely associated is dazzling to think about. The natural assumption is that a person who is very good at processing melody can be expected to be good at imagining a three dimensional arrangement of objects. In today's world, that's interesting. But thinking 10,000 or 11,000 years ago, it was monumental. This information could be giving us some insight about the folks who set wheels in motion for changes that have shaped today's world. Advanced visuospatial skills would have been essential for creating a megalithic structure 10,000 or 11,000 years ago. Doubtless, without it the result would have been a chaotic pile of very big stones.

Obviously, there is room for exploration of this concept, by experts like Drs. Zatorre and Zangenehpour who speak the language. "Several lines of evidence suggest that exposure to only one component of typically audiovisual events can lead to crossmodal cortical activation. These effects are likely explained by long-term associations formed between the auditory and visual components of such events. It is not certain whether such crossmodal recruitment can occur in the absence of explicit conditioning, semantic factors, or long-term association; nor is it clear whether primary sensory cortices can be recruited in such paradigms. In the present study we tested the hypothesis that crossmodal cortical recruitment would occur even after a brief exposure to bimodal stimuli without semantic association. We used positron emission tomography, and an apparatus allowing presentation of spatially and temporally congruous audiovisual stimuli (noise bursts and light flashes). When presented with only the auditory or visual components of the bimodal stimuli, naïve subjects showed only modality-specific cortical activation, as expected. However, subjects who had previously been exposed to the audiovisual stimuli showed increased cerebral blood flow in the primary visual cortex when presented with sounds alone. Functional connectivity analysis suggested that the auditory cortex was the source of visual cortex activity. This crossmodal activation appears to be the result of implicit associations of the two stimuli, likely driven by their spatiotemporal characteristics; it was observed after a relatively short period of exposure (approximately 45 min), and lasted for a relatively long period after the initial exposure (approximately 1 day). The findings indicate that auditory and visual cortices interact with one another to a larger degree than typically assumed."[51]

The implications of all this are fascinating. When we bring modern neuroscience to the oldest man-made sites of the ancient world and mix them with music, something is suddenly making sense.

Without acoustics, archaeology is deaf.
Without archaeology, acoustics is blind,
and "acoustic archaeology" is merely acoustics.
Clearly, convergent facts from both fields
are needed for an investigation
to be designated as acoustic archaeology.

Dr. David Lubman, Researcher and acoustics expert

Chapter 7

The Present of Archaeoacoustics

In this study we have been able to see the obvious route of a cultural tradition of building, of agriculture and of forward development. We know where civilization came from. We are just beginning to understand why. Now, what are we going to do with the knowing?

7.1 Pseudoscience and the Wall of Resistance

Although few will bring it out into the open, a significant finding from the OTSF Archaeoacoustics conferences was the clear division between hard science and "New Age" esoteric practice. There was a genuine fear on the part of the traditional scholars that a valid field of study was going to be impossible because it would never find necessary academic support. The reason was the threat of serious drift from the realm of credible evidence. This is a very real concern.

These conferences were highly unusual gatherings. As has previously been stated, they were transdisciplinary. We felt that since we were collecting a body of knowledge that had never been examined in depth, all voices with something to add should be heard. Included were presentations from independent researchers who had information to share that "rocked the boat" so to speak. Some were seriously worth considering; some did not exactly live up to their abstracts. Nevertheless, we had to listen. It was a great learning experience.

Now, as we move forward, it is the challenge of all of us who participated and those who have read the proceedings to figure out what to make of it. Once a person zeros in on the fact that there is something important about human physical response to special sound in the world's oldest monumental spaces, there comes a drive to reclaim it and apply what we have learned from antiquity to help people in the modern world.

It is the suggestion of the author that this ancient tradition of primitive holistic knowledge did not entirely disappear, but has come down to us today in forms that deviated with the divisions in learning that came with development of scientific methodology and advanced education. Once one has tuned in with some perspective, it is not difficult to recognize that the ancient lore of sound evolved differently within the various population groups as they moved and developed over millennia. A prime example is the intentionality of boosted acoustic properties in historic places of worship.

For the scientist, some of those "evolved" forms can be uncomfortable for the same reason that most archaeologists are not taught to consider sound. The properties are difficult, if not impossible, to measure. That leaves the door open to conjecture that can "go off the deep end", inviting a reflexive resistance to serious study. With a bit of mutual respect, perhaps there is room to bridge this wall and make full use of new learning on both sides of it. The very primitive nature of the origins of this learning can be said to transcend the dictates of the political, religious and academic filters that have developed since history began. Each person can absorb what matters to them and there is no need to worry about what doesn't. As with a radio station that plays only one type of music that doesn't suit, we can change the channel, as it were.

7.2 Research

What we are learning about sound has tremendous potential for benefit in our present. Intonation therapy and other applications in the health industry are abundant. Conditions treated include anxiety, high blood pressure, depression, and autism. According to Alt MD (alternative medicine,) chanting and overtone chanting are used in therapy with Alzheimer's patients because this form of sound therapy is said to help with memory function. Some researchers think that music memories may outlast some other types of memories because music involves so many parts of the brain. There are methods in use for conditions including dyslexia, attention deficit hyperactivity disorder (ADHD), Down syndrome, chronic fatigue syndrome, autism, depression, and behavioral problems.[52]

There is certainly more to be gained from more research into how the brain might be affected by sound at the level of a bass or baritone voice: that level which achieves high impact; resonance naturally produced within megalithic soundboxes like Newgrange, Camster Cairn and Malta's Hal Saflieni Hypogeum. The human brain has not changed a lot in 5,000 years, so we can expect to get a good idea of the impact of that sound on the original people who were using these spaces. Of course, scientific research requires that the stimulation can be controlled for pitch and be repeatable in multiple applications. The first thing a modern researcher might think is that the surest way to get that is with an artificial electronic tone. At the time of this writing, laboratory testing has been conducted only with electronically generated signals. It will be an enlightening day when testing can be done on exposure to the full stimulation of live human voice with all its overtones and nuances. We are inching toward that ultimate collection of data. Nevertheless, what we have is a start.

The value of reclaiming archaeoacoustic awareness as a therapeutic tool is in its infancy. Yet, the very "primitive" quality of its delivery may actually be its strongest endowment. Prof. Reznikoff has worked intensively with the voice and has developed a practice of sound healing in which he has been involved since 1980. He points out that "Our consciousness is structured by the human voice, even before birth, so the voice penetrates very deeply into consciousness, especially when natural resonance is used."

The auditory system of the child in its mother's womb is already formed at six months of pregnancy. As it is developing the fetus hears the lower tones of father's voice before the mother's. Reznikoff goes on to explain that since the developing senses are restricted in utero, the deepest prenatal perceptions of the child are auditory. In therapy, sounds at this "megalithic" level are recognized unconsciously because, "The main harmonics were already heard in the womb as transmitted by the mother's bone structure down to the lower part of her back and the pelvis so that the child is surrounded by them and its early consciousness is impregnated by the fundamental harmonic consonances."

When resources are committed to a serious investigation, it is sincerely hoped that extended research might also be taken up in collaboration with the OTS Foundation for Neolithic Studies, and its recordings of actual vocals and drums vibrating through the Hal Saflieni Hypogeum. At the time of this writing, we are ready to leap into pilot studies with an existing audio/visual prototype application platform that utilizes these recordings of specific sound with all the acoustic characteristics of the space. Target areas for therapeutic value include treatment of depression, Alzheimer's and other dementias, social and behavioral disorders cited above, as well as a meditation tool useful in recovery from addiction.

Still on the subject of research, it is best to forget about ever finding actual recorded sound from antiquity. A cleverly written blog cites some wishful thinking. "Belgian researchers have been able to use computer scans of the grooves in 6,500-year-old pottery to extract sounds—including talking and laughter—made by the vibrations of the tools used to make the pottery." There was a video with interviews in French, but it was said that the pottery sounds could be heard. Again, from the blog: "Voices from six millennia ago! Isn't that amazing? I imagine this sort of recording would be a real boon for historical linguists . . . if the video weren't a hoax."

The writer of this blog did his homework. "Well, hoax might be a little strong. According to a follow-up message, the video was originally an April Fools' joke. . . . You have to admit it's pretty funny. It's not particularly original, though. There are at least three prior instances of this idea (ancient audio recordings in pottery) being used in science fiction. One was in a 1979 short story by Gregory Benford called Time Shards, in which a researcher tries to recover thousand-year-old sound from a piece of pottery thrown on a wheel and inscribed with a fine wire. The second was Rudy Rucker's 1981 story "Buzz", which includes a snippet of similar ancient audio recovered from ancient Egyptian pottery. The third was an episode of The X-Files (television series) titled "Hollywood A.D. in which Mulder and Scully think they may have stumbled across the 'Lazarus Bowl', in which the voice of Jesus is recorded in the same way." This is not proposed as a good line of future research.

Dr. Yioutsos, however, has some excellent suggestions for more research. "In order to discover the 'mindscape' of these ancient landscapes," he writes, "we need to think in multi-sensory terms. Instead of focusing on how ancient people viewed these sacred caves, we should explore how caves were perceived and experienced by them. Therefore, these cave environments should be re-visited, re-mapped and all their visual and aural features should be recorded using sophisticated electronic instrumentation. Furthermore, in an attempt to comprehend how Pan Rituals were experienced by the participants, we should start making noise! The reconstruction of the ancient musical instruments used in these ceremonies (rattles, double flutes, syrinx) and the recording of their sound performed inside these holy caverns in conjunction to various human sound effects (laughing, yelling, singing, dancing, clapping) would be helpful in order to understand how sound reacted to the cave environment and possibly how it affected the human sensorium leading occasionally the human mind to altered states of consciousness."

The field of genetics doesn't need a new suggestion, but maybe in collaboration with neuroscience and therapeutic approaches, there is something. If it was possible to make the biological change for digesting milk, one must wonder if a brain with a high capacity for processing both music and spacial relationships went with it. In 2001, geneticists looking for the root of a speech problem tracked it down to a mutation in a gene they named FOXP2, the so-called "Language Gene." It will probably be geneticists who identify a mutation in the gene for music processing and spacial visualization who could give us confirmation.

7.3 Application

As for practical use of what we have learned so far, there are some further ideas.

To date, there is no university program preparing students for a degree in the field of Archaeoacoustics. It would be very interesting if a course of study could be prepared and offered, at least for students in archaeology and anthropology. There should be some way to accommodate people who want to study in more depth and/or actually participate in a site evaluation.

We can make more music together. Historically, it has always been very common for people to come together to make or listen to music at the same time. When people do this, there is a tendency to share a similar emotional state, to sense a real connection to the people around them. The bonding process that results is a giant step toward reversing the isolation and detachment that seem to be developing in our Western society that is so saturated in electronic entertainment. There is every indication that group musical behavior translates into more cooperative and prosocial conduct. We can support and insist on music programs in our schools. This is a sad area to make a sacrifice for budget cuts when we understand how much it can do for our children.

We are hearing from the Wellness community: "How can we create spaces (or use existing spaces) where people can feel safe to share and to receive sounds that connect us on a tribal/primal level. I feel if we can connect more as humans in a deeper, more felt way, we can overcome many of our problems." It's a great idea. Kiosks in every noisy civilized city would be good.

Perhaps we can encourage architects to include some quiet contemplative spaces in the noisy urban population centers that are blooming across the world. The American Institute of Architects already has guidelines and there are strict regulations for noise levels in hospital nurseries these days, in response to a study of stress hormone levels in infants.

Figure 7.1

Toward this end, a major project has become the development of a Sonic Exploratorium Nodal Structure, as introduced by architect Shea Michael Trahan in the Bonus Material of this book. (Project updates at OTSF.org) Exploiting the symmetrical nature of cymatic pattern, scaled to a precise multiple of the desired tone, this structure becomes a powerful expression of sound in architecture as the room becomes an instrument. The intention is for this structure to function across several arenas:

- For research—a controlled environment for an extended study of neurological impact of certain resonant sound on brain activity
- For therapy—a multi-sensory tool for treatment of addiction, depression and social disorders, as well as memory and cognitive disorders
- For wellness—a space where people can feel safe to share and to receive sounds that impress and connect us on a primal level, as well as practice self-guided and structured meditation
- For education—an experiential mini-museum for student groups and a curious public to learn something about the physics of sound, its roll in human development and the advent of Western civilization

It does seem that the archaeology of sound, as we have pursued it in this book, has something to say to people living in the 21st century. Will they hear it?

References

1. Donald, M. (March 2009), The sapient paradox: Can cognitive neuroscience solve it?, *Brain*, Volume 132, Issue 3, pp. 820–824, https://doi.org/10.1093/brain/awn290.
2. Watson, A. (n/dated), Archaeoacoustics, http://www.aaronwatson.co.uk, accessed on 20 June 2016.
3. See the Heritage Malta website for images, full descriptions and information about tickets. http://heritagemalta.org/museums-sites/hal-saflieni-hypogeum/.
4. *National Geographic Magazine* (1920), Volume XXXVII, Issue 5, p. 471.
5. See https://www.scienceworld.ca/resource/sound-vibration-vibration-vibration/, accessed 23 August 2023.
6. Britannica (17 July 2023), the Editors of Encyclopaedia. "Resonance". *Encyclopedia Britannica*, https://www.britannica.com/science/resonance-vibration, accessed 3 August 2023.
7. Kreisberg, G. (2014), *Proceedings of Conference "Archaeoacoustics: The Archaeology of Sound"*, Malta, 19–22 February, pp. 267–268.
8. Devereux, P., and Jahn, R. G. (1996), Preliminary investigations and cognitive considerations of the acoustical resonances of selected archaeological sites, *Antiquity*, Volume 70, Issue 269, pp. 665–666.
9. Cook, I. A., Pajot, S. K., and Leuchter, A. F. (2008), Ancient architectural acoustic resonance patterns and regional brain activity, *Time and Mind*, Volume 1, Issue 1, pp. 95–104.
10. Debertolis, P., Tirelli, G., and Monti, F. (2014), Systems of acoustic resonance in ancient sites and related brain activity, *Proceedings of Conference "Archaeoacoustics: The Archaeology of Sound"*, Malta, 19–22 February, pp. 59–65.
11. Kreisberg, G. (2014), *Proceedings of Conference "Archaeoacoustics: The Archaeology of Sound"*, Malta, 19–22 February, pp. 267–268.

12. Lindstrom, T. C., and Zubrow, E. (2014), Fear and amazement, *Proceedings of Conference "Archaeoacoustics: The Archaeology of Sound", Malta*, 19–22 February, pp. 255–264.
13. Stroud, K. (2014), Hal Saflieni Hypogeum—acoustic myths and science, *Proceedings of Conference "Archaeoacoustics: The Archaeology of Sound", Malta*, 19–22 February, pp. 37–43.
14. Walter R. (1940), Wanderers awheel in Malta, *National Geographic Magazine*, Volume 78, Issue 2, pp. 253–272.
15. See https://www.popsci.com/science/article/2012-10/eating-cooked-food-made-us-human-study-finds/, accessed 13 August 2023.
16. See https://www.smithsonianmag.com/history/gobekli-tepe-the-worlds-first-temple-83613665/, accessed 4 August 2023.
17. Bollongino, R., et al. (2012), Modern taurine cattle descended from small number of near-eastern founders, *Molecular Biology and Evolution*, Volume 29, Issue 9, September 2012, pp. 2101–2104.
18. Arunn Murugesu, J. (2023), See https://www.newscientist.com/article/2385057-origin-of-indo-european-languages-traced-back-to-8000-years-ago/, accessed 4 August 2023.
19. Wang, K., et al. (2023), High-coverage genome of the Tyrolean Iceman reveals unusually high Anatolian farmer ancestry, *Cell Genomics*, 16 August, 2023. doi: 10.1016/j.xgen.2023.100377.
20. See https://www.bbc.com/news/science-environment-47938188, accessed 4 August 2023.
21. See https://www.dailymail.co.uk/sciencetech/article-6826531/Ancient-Babylonian-treasure-seized-Heathrow-airport-returned-Iraq.html, accessed 25 August 2023.
22. Schmidt, K. (2011), Gobekli Tepe. *The Neolithic in Turkey*, Volume 2, Archaeology & Art Publications, Istanbul, pp. 41–83.
23. England, R. (2014), Neolithic architecture—space and sound, *Proceedings of Conference "Archaeoacoustics: The Archaeology of Sound", Malta*, 19–22 February, pp. 33–36.
24. Christian, L. (2023), personal correspondence, www.Academia.edu, 13 AUG 2023.

25. Till, R. (2014), *Proceedings of Conference "Archaeoacoustics: The Archaeology of Sound", Malta*, 19–22 February, pp. 23–32.

26. Reznikoff, I. (2014), *Proceedings of Conference "Archaeoacoustics: The Archaeology of Sound", Malta*, 19–22 February, pp. 45–50.

27. Yioutsos, N. P. (2014), Pan rituals of ancient Greece: A multisensory body experience, *Proceedings of Conference "Archaeoacoustics: The Archaeology of Sound", Malta*, 19–22 February, pp. 67–78.

28. See https://ich.unesco.org/en/RL/grand-song-of-the-dong-ethnic-group-00202, accessed 18 August 2023.

29. Conner, S. (2014), The score of Babylon—outline of an interdisciplinary framework of reconstructing Mesopotamian song, *Proceedings of Conference "Archaeoacoustics: The Archaeology of Sound", Malta*, 19–22 February, pp. 195–209.

30. Lovász, I. (2023), see https://lovasziren.hu/en/news/healing-voice/, accessed 22 September 2023.

31. Lewis-Williams, D., and Pierce, D. (2005), *Inside the Neolithic Mind: Consciousness, Cosmos and the Realm of the Gods,* Thames and Hudson, London.

32. See https://en.wikipedia.org/wiki/Inside_the_Neolithic_Mind, accessed 16 August 2023.

33. Trump, D. H. (2005), *Malta Prehistory and Temples*, Midsea Books, Ltd., Malta, p. 118.

34. Devereux, P. (2001), *Stone Age Soundtracks*, Vega/Chrysalis Books, London, pp. 91–92.

35. Evans, J. D. (1971), *The Prehistoric Antiquities of the Maltese Islands: A Survey,* The Athlone Press, London.

36. Reznikoff, I. (2014), The Hal Saflieni Hypogeum: A link between paleolithic painted caves and romanesque chapels?, *Proceedings of Conference "Archaeoacoustics: The Archaeology of Sound", Malta*, 19–22 February, pp. 45–50.

37. Coimbra, F. (2014), An interdisciplinary approach: The contribution of rock art for archaeoacoustic studies, *Proceedings of Conference "Archaeoacoustics: The Archaeology of Sound", Malta*, 19–22 February, pp. 51–58.

38. Schmidt, K. (2012), personal communication.
39. See https://reneefleming.com/advocacy/music-and-the-mind, accessed 8 July 2021.
40. See https://neuro.hms.harvard.edu/centers-and-initiatives/harvard-mahoney-neuroscience-institute/about-hmni/archive-brain-1, accessed 9 July 2021.
41. See https://www.didjshop.com, accessed 20 June 2016.
42. Xu, J., Vik, A., Groote, I. R., Lagopoulos, J., Holen, A., Ellingsen, Ø., Håberg, A. K., and Davanger, S. (2014), Nondirective meditation activates default mode network and areas associated with memory retrieval and emotional processing. *Frontiers in Human Neuroscience*, Volume 8, pp. 1–10.
43. Salimpoor, V., Benovoy, M., Larcher, K., Dagher, A., and Zatorre, R. J. (2011). Anatomically distinct dopamine release during anticipation and experience of peak emotion to music. *Nature Neuroscience*, Volume 14, pp. 257–262, doi:10.1038/nn.2726.
44. See https://www.webmd.com/mental-health/what-is-dopamine #1-2, accessed 14 July, 2022.
45. See https://hms.harvard.edu/news-events/publications-archive/brain/music-brain, accessed 5 August 2023; https://www.sciencerepository.org/music-and-dopamine_JBN-2020-1-102, accessed 5 August 2023.
46. Blood, A. J., and Zatorre, R. J. (2001), Intensely pleasurable responses to music correlate with activity in brain regions implicated with reward and emotion. *Proceedings of the National Academy of Sciences*, Volume 98, pp. 11818–11823.
47. See https://health.clevelandclinic.org/is-addiction-genetic, accessed 15 August 2023.
48. Delestraty, C. (2014), http://thoughtcatalog.com/cody-delistraty/2014/01/music-as-religious-experience-the-neuroscience-of-a-song, accessed 21 June 2016.
49. Newberg, A., and D'Aquili, E. (1999), *The Mystical Mind: Probing the Biology of Religious Experiences*, Fortress Press, Minneapolis, MN.
50. See https://www.challenging-our-minds.com/blog/?p=38, accessed 28 June 2016.

51. Zangenehpour S., and Zatorre R. J. (2010), Crossmodal recruitment of primary visual cortex following brief exposure to bimodal audiovisual stimuli. *Neuropsychologia*. 2010, Volume 48, Issue 2, 591–600. doi: 10.1016/j.neuropsychologia.2009.10.022.

52. A good list of resources about sound in medicine can be found on the AltMd website: http://www.altmd.com/Articles/Sound-Therapy--Encyclopedia-of-Alternative-Medicine, accessed 22 June 2016.

BONUS MATERIAL

Reprinted here by permission of The OTS Foundation for Neolithic Studies are five select papers from the conferences on Archaeoacoustics. The full collection of papers from the Malta inaugural meeting as well as the proceedings of subsequent conferences in Istanbul, Turkey and Tomar, Portugal are published independently in separate volumes.

Resonant Form: The Convergence of Sound and Space

Shea Michael Trahan

SHEA MICHAEL TRAHAN, AIA, LEED AP, is a licensed architect and an Associate within the New Orleans-based firm of Trapolin-Peer Architects where he directs evidence-based design initiatives. Shea holds degrees in architecture from the University of Louisiana at Lafayette and Tulane University as well as a certificate in Neuroscience for Architecture from the Newschool of Architecture and Design. His research combines aspects of architectural acoustics, neuroscience, and algorithmic design and has been featured in a TEDx presentation entitled *The Architecture of Sound*, an exhibit at the New Orleans Museum of Art, a poster presentation at the 2016 Academy of Neuroscience for Architecture conference, and published within the textbook *Creating Sensory Spaces: The Architecture of the Invisible*.

Abstract

The built environment is a powerful tool for affecting human awareness by embodying experiences which interact with biological rhythms to shift states of consciousness. Through exploration of precedents of architectural sonic phenomenon from throughout history, this body of research aims to identify powerful sonic tools towards such an affect. Looking to the future of designing such spaces, the research delves into algorithmic design and

---

Megaliths, Music, and the Mind
Linda C. Eneix
Copyright © 2024 Jenny Stanford Publishing Pte. Ltd.
ISBN 978-981-5129-25-0 (Hardcover), 978-1-003-46823-3 (eBook)
www.jennystanford.com

cymatic processes to seek to create forms which are embodiments of sound.

Keywords: Neuroscience, cymatics, algorithm.

"Listen! Interiors are like large instruments, collecting sound, amplifying it, transmitting it elsewhere."
Peter Zumthor.[1]

Human spatial perception is a sensual experience of the world we inhabit. While we experience architecture through all of our senses by varying degrees, the process of design has long preferenced the visual at the expense of other modes of perception. This body of research and proposed design methodology aims to focus on acoustic aesthetics to create spaces which manifest sonic phenomenon that not only elicit psychological responses for inhabitants but also induce physiological shifts toward meditative and/or mystical states. As humans are sonic instruments themselves through use of our hearing and vocal range, the engagement of our sonic qualities within a sensitively designed aural architecture creates the potential for a truly transcendent and immaterial experience. It is in this way that this design proposal strives to enlist the resonant natures of architectural forms to deeply engage and expand the sensory awareness of human spatial perception.

The world which we inhabit is a symphony of oscillating systems, manifesting themselves through a variety of scales. Daily, monthly, and annual cycles are all examples of oscillating systems which impact human symbiosis through their frequency and intensity. Interestingly, the human body is itself

---

[1]Zumthor, P. *Atmospheres: Architectural Environments, Surrounding Objects* (Basel; Boston: Birkhauser, 2006): 29.

asymphony of oscillating systems, each carrying out vital rhythms to maintain life, health, and consciousness. Cardio-respiratory rhythms regulate breathing, heart rate, and blood pressure. Pedestrians walking together find that their strides may synchronize in the course of a stroll. The brain conducts its own activities via neural oscillations, or brainwaves.

This inherently rhythmic aspect of human existence permeates the way people experience and interact with the world around them. A characteristic of oscillating systems which is critical to this experiential interaction is the process of entrainment. Through entrainment, two oscillating systems may interact with one another and eventually come to synchronize to a singular frequency.[2] In the case of circadian rhythms, the human body seeks entrainment from the rhythm of the Earth's revolutions.

Entrainment is not specific to biological systems and may be observed in mechanical systems. Non-synchronized metronomes placed on a flexible base will eventually exchange enough interaction to shift phase and become synchronous. However, the fact that a biological rhythm may be definitively changed by an outside frequency is critical to our consideration of architecture. Considering that the EPA estimates that Americans spend up to 93% of their time indoors,[3] the quality and characteristics of the spaces we inhabit are very important for wellbeing, even to the extent of shifting biological functions. A team of designers at the Mixed Reality Lab Nottingham showed exactly this in their 2010 installation and experiment entitled ExoBuilding. In the biofeedback installation, participants spent extended durations seated beneath a tensile fabric canopy

---

[2]Néda Z, Ravasz E, Brechet Y, Vicsek T, Barabsi AL (2000). Self-organizing process: The sound of many hands clapping, *Nature*. 403: 849–850.

[3]Klepeis N, Nelson W, Ott W, Robinson J, Tsang A, Switzer P, Behar J, Hern S, Engelmann W (2001). "The National Human Activity Pattern Survey (NHAPS): a resource for assessing exposure to environmental pollutants". *Journal of Exposure Analysis and Environmental Epidemiology*. 11: 231–252.

which stretched away from them as they inhaled and collapsed back near them as they exhaled. The participant's heart rate was also played as a real-time audio track through a subwoofer. What the experiment showed was that participants were able to reach a calmed state similar to mindfulness meditation through the experience of the immersive environment with slower, deeper breathing occurring after a few minutes of experience. Even more compelling was a second experiment in which the participants were similarly introduced to the immersive space as before, but soon after beginning the experiment the biofeedback software took over the rate of movement of the canopy, slowing it by 20%. The result was that a majority of participants entrained their respiration to match the slower rate of the installation, and were thus brought to the calmed state quicker by environmental input.[4]

Such entrainment can happen in both the brain (brainwave synchronization) and the body (cardiorespiratory coupling), and may be induced by rhythmic visual or auditory stimuli. Indeed, the use of sound for such effects is a powerful tool. The sonic environment has a drastic impact on human wellbeing partially due to the fact that the human brain is hardwired to constantly evaluate the aural landscape. The ear enjoys three times more nerve connections to the brain than the eye does[5] and can decipher a range of sound from 20 Hz to 20,000 Hz.[6] This range of frequency is an order of magnitude of 1000 times, equivalent to roughly 10 octaves of sound. Considering the range of frequencies detectable by the eye, it equates to an order of magnitude of only 2, or a single octave of frequency, experienced as colors rather than as tones. We know this 'octave' of color as ROYGBIV.

---

[4]Schnadelbach, H. "Embodied adaptive architecture." *YouTube*, 24 September 2016, https://youtu.be/HEkFjx4kxmc.
[5]Crowe B., *Music and Soulmaking: Toward a New Theory of Music Therapy*, Scarecrow Press, 2004.
[6]Helmholtz H., *On the Sensations of Tone*, Dover Publications, 1954.

As exhibited, humans are greatly affected by the physical (and specifically sonic) environment. Given the predominance of the built environment in modern human experience, the design of our spaces is coming to be seen as a powerful tool for human wellbeing. Architecture is the setting in which we experience play, love, rest, learning, struggle, adaptation, and even transcendence. Could architecture go beyond simply containing experience and instead act as a catalyst for transcendent experience?

In search for such a sonic gateway to transcendent experience, two characteristics of spatial sound come to the forefront; reverberation and resonance. Reverberation shall be discussed as the period of time a sound takes to dissipate within a space, while resonance may be understood as the predominant frequency at which a space or object vibrates.

To experience magnificent examples of reverberation one only needs to visit a gothic cathedral, or perhaps the churches in which Gregorian chant developed. A specific structure is the Baptistry of St. John adjacent to Pisa's iconic Leaning Tower. The Romanesque/Gothic baptistry manifests a rather unique form of reverberation imparting the space with a powerful sonic tool.

Most spaces have reverberation times of a fraction-of-a-second, meaning that once a sound is discontinued the energy of the source sound disappears almost instantaneously. This short moment of continued sound is created by the original sound waves bouncing between the walls of the space until they are eventually absorbed by the materials within the room. The more reflective the materials, the more bounces which are possible (this is the commonly known effect enjoyed by those who sing in the shower). The shape and arrangement of a building also affects the bouncing of waves within a space and can either diffract, focus, delay, or redirect the bouncing sound. While most structures have minuscule reverberation times, many churches and concert venues can approach reverberation times of two seconds.

Due to the particular form and material at Pisa, when a vocalist sings a tone within the Baptistry they are greeted by a reverberation time many times longer than the grandest of concert halls. This allows one to layer their vocal energy, even harmonizing with their own voice as it reverberates within the space for up to 12 seconds. The vocal energies of the singer slowly fill the building's large volume, held in momentary suspension by the geometry of the structure before returning to the listener from another direction.

Studying the Baptistry form and construction, you find that the marble used to construct the space is highly reflective thus bouncing most of the energy from a given sound wave back towards the interior of the room. Working in tandem with this strong reflectance is the arrangement of the walls, columns, and roofs. In plan the Baptistry is a double-layered circle with an exterior diameter of one hundred and sixteen feet. The circular arrangement guarantees that any sound within the space will travel outward radially only to be reflected and re-focused back toward the interior of the room. This refocusing conserves the sonic energy by maximizing the amount of sound returned to the listener.

The interior columns which carry the weight of the gallery and ceiling above play a crucial role in the production of the sonic environment as well. Sound waves may take one of three radial paths in plan. Sound waves may be reflected by the interior face of the column and sent back toward the center of the space. This represents the shortest path back to the center and thus it is the fastest. Some of the sound waves will miss the columns and travel to the exterior wall before being reflected back to the interior of the space. These take a slightly longer path (in both distance and time). A third wave path travels past the columns, is reflected by the exterior walls, but rather than returning directly to the interior of the space, these waves are

The Sapient Development Timeline | 183

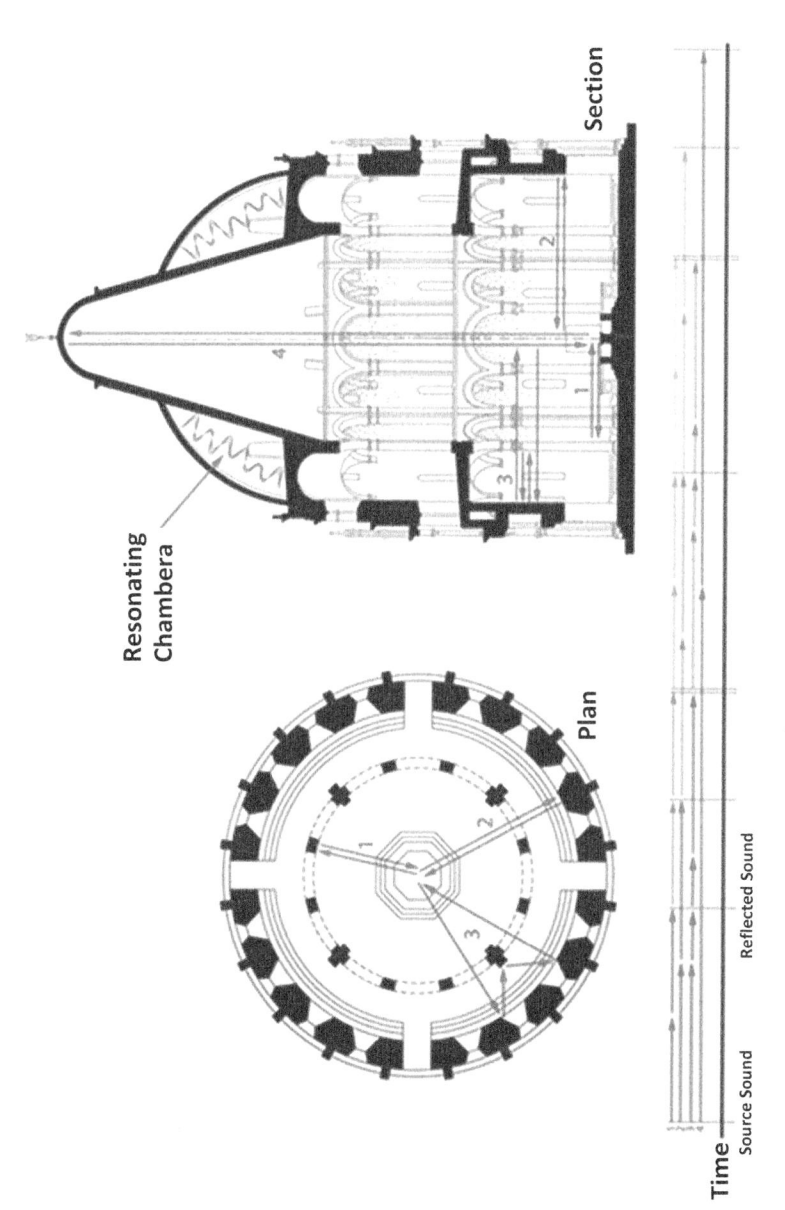

Figure 1 Sonic analysis diagram of Pisa Baptistry.

again bounced by the exterior of the column. In this moment the sound wave is deflected back and forth between the column and outer wall, further extending the time between the source sound and its deflected return to the center of the room. In this way the architecture manipulates the acoustic paths of sound waves creating a layering of reflections and a dramatically extended reverberation of the sound.

By studying the Baptistry in section, one can see the original roof (observable as the interior ceiling form) along with the modified design which created a second skin in the shape of a dome around the original roofline. Whether by accident or intent, this double layered spatial arrangement creates a resonating chamber between the two skins. These architectural traits combine to intensify the sonic character of the space.

Arguably the most focused architectural example of resonance is to be found within the Hypogeum Hal Saflieni in Paola, Malta. This prehistoric archaeological site contains an underground temple discovered in 1902. A noteworthy element of the temple is a room called the Oracle Chamber. This space is oval shaped in both plan and section with ridges carved into the ceiling of roughly seven feet in height. The ceiling is particularly important as it is elaborately painted in a swirling red paint, potentially a musical annotation. Also notable about the chamber is the presence of a small portal which allows sounds from within the Oracle Chamber to be transmitted and clearly audible to all the rooms of the complex.

Acoustic analysis of the Oracle Room has identified a resonant frequency of 110 Hz.[7] Standing in this space one is able

---

[7] Cook I., Pajot S., Leuchter A., *Ancient Architectural Acoustic Resonance Patterns and Regional Brain Activity*, in Time and Mind, Volume 1, Number 1, 2008. [Note from Linda Eneix: We now know that the number 110 Hz is not singularly fixed. The acoustics of Hal Saflieni are extremely complex, with more than one frequency triggering high resonance response in the spaces, most lying between 60 Hz and 120 Hz. Thus the chance for any physiological impact is increased, since sensitivity seems also to vary somewhat from one person to another and even at different times in the same individual.]

The Convergence of Sound and Space | 185

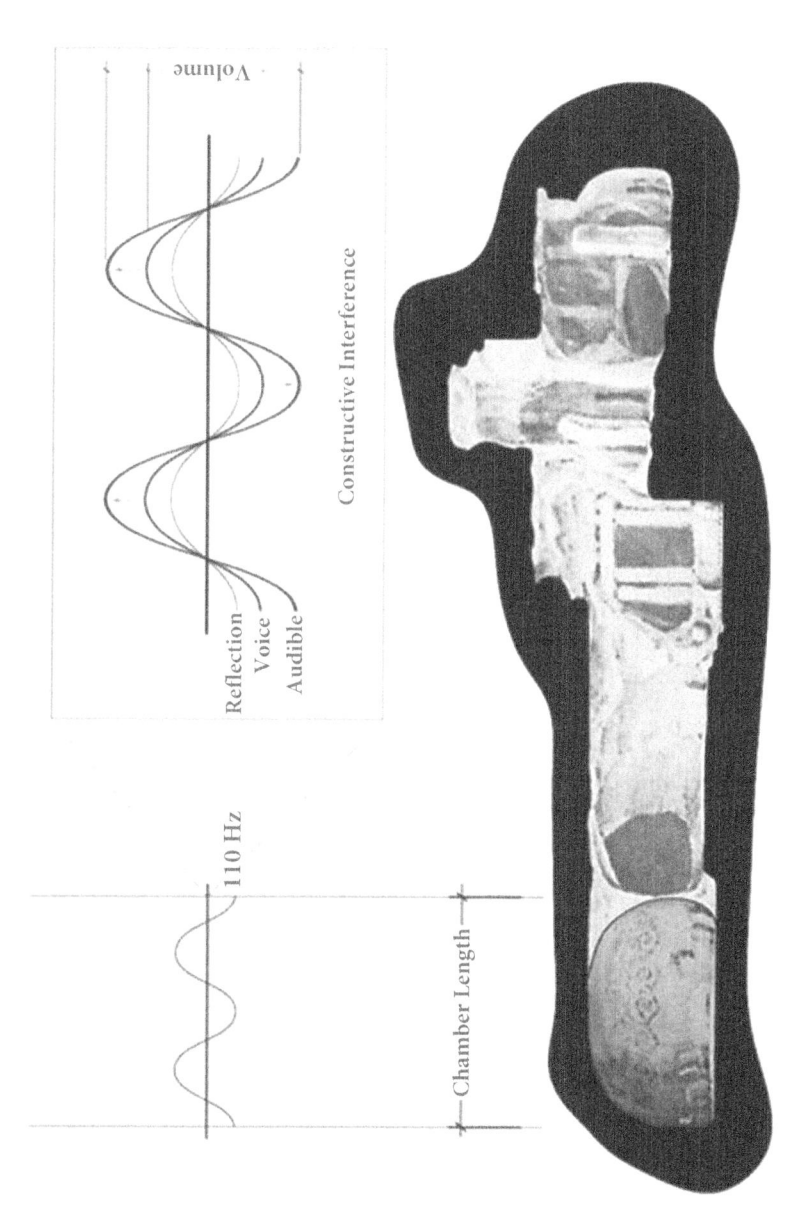

Figure 2 Sonic analysis diagram of Hypogeum Ħal Saflieni.

to identify the resonant frequency of the room by singing or humming at the lowest range of your vocal register and slowly raising the pitch of the tone. When the resonant frequency is reached at 110 Hz, a multisensory event takes place. The intensity of the sound of your voice within the space increases instantaneously, sounding as though someone has joined in singing. This is caused through constructive interference, created when the sound waves of your voice synchronize. This is a natural form of amplification (Fig. 2). Additionally, the character of the sound becomes increasingly spatial, observably approaching you from all directions simultaneously and taking on a vibrotactile quality so that you sense the sound as a tingling vibration through the skin. When you discontinue singing the resonant tone, it remains notably audible for a few brief moments before decaying into silence. The experience is at once disorienting (you feel as though your equilibrium is lost) and awe inspiring. This sonic interaction—by which the space informs the performer—inverts the relationship between musician and instrument in that it is indeed the instrument which tunes the singer.

Judged upon the characteristics of the acoustic effects alone, the Hypogeum is an impressive site. More impressive though are the results of a study into the effects of the Hypogeum's particular frequency, 110 Hz, upon the human brain. When the brain is exposed to the resonant frequency of the Hypogeum, the tone causes a shift in the prefrontal cortex from left dominance to right dominance.[8] This shift de-activates the language centers of the brain (focused on rhythm and patterns) and hyper-activates an emotional center often associated with the perception of experiences of spirituality and enlightenment.

This type of physiological reaction to an environment is precisely the sort of entrainment discussed earlier. Indeed, it appears that the study of neuroscience as it relates to

---
[8]Ibid.

architecture has become the forefront of this realm of research. Given the strong connection between the sense of hearing and the brain, the sonic realm promises a wealth of potential for creating even more powerful sensory expanding spaces.

The human brain interprets sounds in a variety of ways. Rhythmic sounds of tapping and drumming lead to activation within the left frontal and left parietal lobes of the brain.[9] Tones activate the brain in an entirely different way, activating areas within the right prefrontal cortex which are associated with emotional states and somatosensory perception.

The same areas of the brain which are involved in the perception of sound also play an important role in experiences of enlightenment, or spiritual awareness. In his 2016 book entitled *How Enlightenment Changes the Brain*, Dr. Andrew Newberg reviews decades of brain scan studies conducted on experienced Buddhist meditators, Franciscan nuns, and Sufi mystics. The recurrent results from his and other studies is that at the moment of enlightenment these spiritual practitioners experience a rapid and significant decrease in the frontal and parietal lobes of the brain.[10] This decrease in regional activity within the brain appears to be most heavily felt in the left hemisphere of the frontal and parietal lobes.

For a listener experiencing sound within an immersive environment such as the Hypogeum or the Pisa Baptistry, the initial experience of sound would be met with increased activity within various brain regions: first the primary auditory cortex followed by the secondary auditory cortex. From here the brain begins to send signals to both the parietal and frontal lobes. If the sonic event is significantly immersive and of substantial

---

[9]Wilkins A., What happens to your brain under the influence of music, accessed July 17, 2017. http://io9.gizmodo.com/5837976/what-happens-to-your-brain-under-the-influence-of-music.

[10]Newberg A., Waldman M., *How Enlightenment Changes Your Brain*, Penguin Random House, 2016.

duration the brain then shifts with rapid decreases in the activity in those areas. This is consequential as the parietal lobe takes in sensory input and creates one's sense of self. A reduction in activity is thus associated with a loss of one's sense of self, creating a new sense of unity or oneness with the universe, God, nature, or consciousness.[11] This notable increase in self-transcendence associated with the drop in parietal lobe activity casts a new light on the neurobiological basis of altered spiritual and religious attitudes.[12]

As the secondary auditory cortex sends signals to the parietal lobe, so too it sends signals to the frontal lobe. The power of sound on the frontal lobe, specifically via the prefrontal cortex, has already been seen in the Hypogeum studies. As in the parietal lobe, given a stimulus of significant duration and intensity, one may experience a sudden drop in activity, predominantly localized to the left frontal lobe. This particular area of the brain is involved in distinguishing the rhythmic patterns of language and is particularly active during focused purposeful concentration. Decreasing activity in this area is notably associated with a sense of surrender as the control mechanisms of the brain give way.[13] Adding to this experience is an increase in the activity of the anterior cingulate cortex, an area of the brain which regulates emotional responses and communicates directly with the self-critiquing tendencies of the frontal lobe, maintaining a balanced control over emotional reactions. As the frontal lobe's control mechanisms decrease and the anterior cingulate cortex increases, the experience of intense emotional responses becomes increasingly possible.[14]

---

[11]Ibid.
[12]Urgesi C., Aglioti S., Skrap M., Fabbro F., *The Spiritual Brain: Selective Cortical Lesions Modulate Human Self-transcendence*, in Neuron, 2010.
[13]Newberg A., Waldman M., *How Enlightenment Changes Your Brain*, Penguin Random House, 2016.
[14]Ibid.

"If the Doors of perception were cleansed everything would appear to man as it is, Infinite."

William Blake[15]

One may be prone to question the value in such an experience, or dismiss it as some sort of fleeting high. Dr. Andrew Newberg addresses this concern directly by stating that a majority of his study subjects have found, "great meaning and purpose to their life", through such enlightenment experiences. He correlates such experiences to defense against depression and anxiety, and cites a study which found that, "a sense of purpose enhances 'coping, generosity, optimism, humility, mature identity status, and more global personality integration'". In an increasingly secular world, Dr. Newberg calls for the need to, "find new tools and experiences to enlighten the minds of the next generation of seekers."[16] This call for a new form of secular ritual is mirrored by the philosopher Alain de Botton. He views ritual as communal acts intended to build community while mediating between the needs of the individual and those of the group.[17] Botton goes even further to specifically tie such ritual to architectural space and calls for a new "Temple of Perspective" as a tool to help contemporary humans cope with the psychological, emotional, and communal needs which religion has traditionally served. "…No less than the church spires in the skyscapes of medieval Christian towns, these temples would function as reminders of our hopes…they would all be connected through the ancient aspiration of sacred architecture: to place us for a time in a thoughtfully structured three-dimensional space, in order to educate and rebalance our souls."[18]

---

[15] Blake W., *The Marriage of Heaven and Hell*, Oxford University Press, 1975.
[16] Newberg A., Waldman M., *How Enlightenment Changes Your Brain*, Penguin Random House, 2016.
[17] Botton A., *Religion for Atheists*, Vintage International, 2013.
[18] Ibid.

Having explored the scientific power of sound to induce awe and manipulate states of consciousness, and having identified the acoustic phenomenon which might be created through architecture to induce such effects, the question becomes, "Where do we start in the design of such spaces?". It may be argued that in search of design inspiration for spaces intended to affect human experience of frequencies, a designer ought to look to nature's existing vibrational patterns for direction. Specifically related to sound, one branch of physics related to modal vibrational phenomenon is cymatics.

The study of cymatics reveals the reality that sound is a spatial experience, traveling through media, and that such processes are inherently formal in nature. Such forms are revealed in two-dimensional representations known as cymatic patterns. So it is that this author began the search for a space designed to manifest sound by investigating the possibility of using sound (through cymatics) as an active design agent. Beginning with an algorithm which mapped the cymatic patterns of various tones I have explored hundreds of mandala-like representations of sonic manifestation.

Exploiting the symmetrical nature of the cymatic pattern, the algorithm then orbits to explore the spatial implications of the sonic forms in three dimensions (Fig. 4). Scaled to a precise multiple of the desired tone, such forms could become inhabitable spaces which might offer a tunable sonic chamber similar in performance to the Hypogeum Hal Saflieni. Through layered spatial complexity and material reflectance the Nodal Structures could also offer extended reverberation times much akin to the sonic phenomenon within the Pisa Baptistery.

Such a hyper-resonant and hyper-reverberant space could offer a powerful embodiment of sound in architecture; a temple of sound which not only embodies, but is also manifest of, sound. Users could delight in discovering the resonant frequency as they sing out searching for the constructive

The Convergence of Sound and Space | 191

interference as the space comes alive in symphony. Extended durations of singing would offer not only an immersive sonic experience, but would also act to slow the breath, only further

Figure 3 Cymatic explorations of tones.

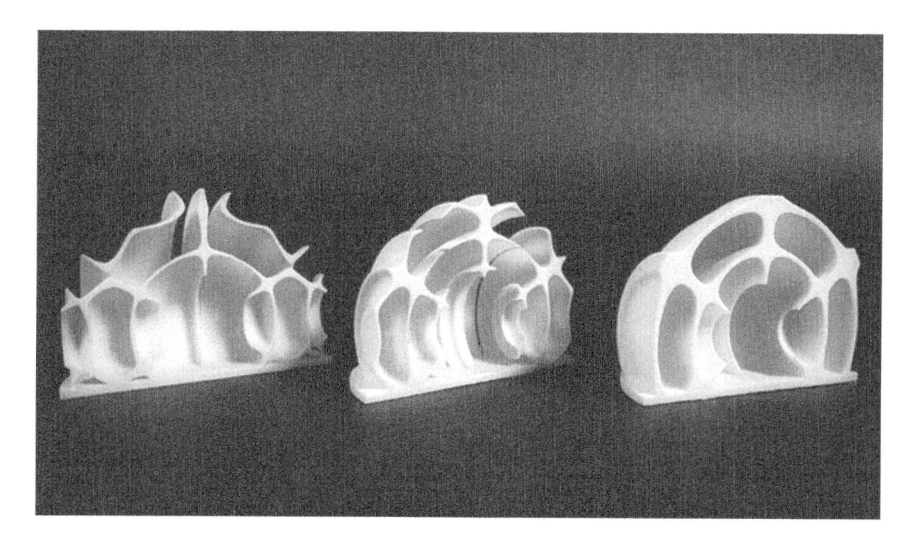

Figure 4 Nodal structures of a minor triad.

embodying the potential for meditative states as the brain areas begin to shift in response to the sensory input (Fig. 5).

Figure 5 Temple of sound.

Figure 6 Sonic laboratory.

As a home for contemporary ritual, the Nodal Structures might offer a new form of secular sacred space, designed to help expand consciousness and prime the brain for enlightenment experiences. Beyond fulfilling the need for such a ritual, these sonic embodiments might offer powerful research opportunities for neuroscientific and physiological experimentation to further explore the power of sound on the human mind and body. Additionally, said space may offer assistance in a variety of sound-based therapies used in treatment for a variety of conditions including depression, PTSD, Alzheimer's, and insomnia (Fig. 6).

The Ħal Saflieni Hypogeum: A Link between Palaeolithic Painted Caves and Romanesque Chapels?

Iegor Reznikoff

IEGOR REZNIKOFF, Emeritus Professor at the University of Paris (Université de Paris Ouest), is a specialist of the foundations of Art and Music of Antiquity and particularly of Early Christian Chant for which his performance, based on deep comparative studies and on practice of ancient scales, is known worldwide. This specialty has led him to study the resonance of Romanesque and Gothic churches as well as the resonance of prehistoric painted caves for which he has shown the relationship between paintings and acoustics. He is a specialist in Sound Therapy and more generally in Sound Anthropology.

(See http://www.musicandmeaning.net/issues/showArticle.php?artID =3.2.)

The Hal Saflieni Hypogeum in Malta is dated, depending on the different parts of it, between 4000 and 2500 BC. Since many human bones and skeletons have been found there, it was certainly intended and used, as a hypogeum is supposed to be: for funeral purposes.[1] The underground depth,

---

[1]See Pace A., *The Hal Saflieni Hypogeum*, Heritage Malta ed., Malta, 2004; or the more complete study, Pace A., *The Hal Saflieni Hypogeum 4000 BC–2000 AD*, Malta, 2000.

Megaliths, Music, and the Mind
Linda C. Eneix
Copyright © 2024 Jenny Stanford Publishing Pte. Ltd.
ISBN 978-981-5129-25-0 (Hardcover), 978-1-003-46823-3 (eBook)
www.jennystanford.com

the architectural achievement, imply the importance of the invisible "after death world" for Neolithic tribes who built the Hal Saflieni Hypogeum. This implies also moments of worship, prayer, devotion: indeed the remarkable small statue, called *The Sleeping Lady* (see Picture 1), was found there as well as many burial ornaments and other figurines. Moreover, the Hypogeum is partially decorated with red geometric ochre paintings (Picture 2).

The Hypogeum appears, therefore, to be a funeral crypt, or chapel. Questions arise about the kind of funeral worship practised there. Since prayer and worship are essentially vocal, the question can be formulated as: *whether it is possible to get some information about the use of the sound of the voice inside the Hypogeum by the worshipping tribes using this underground monument.*

Picture 1 *The Sleeping Lady*, ca. 3600–2500 B.C., National Museum of Archaeology, La Valette, Malta (Photo Malta Heritage).

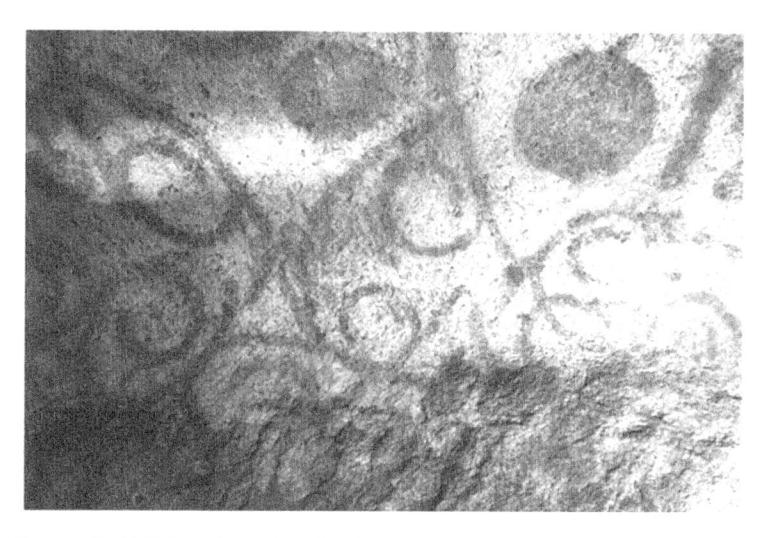

Picture 2 *Wall Paintings*, Oracle Chamber, H.S. Hypogeum, Malta (Photo Malta Heritage).

Keeping in mind the remarkable achievement of the Hypogeum from the architectural point of view (Picture 3) and of the *Sleeping Lady* from the art point of view, it is naïve to think of the chant or music performed there, as being "primitive".

Some ignorant performers tend to recreate ancient music using stones to make *clic-clic*, or shrieking or playing flutes like children who just discover the instrument (with only two holes, a good musician can play the whole chromatic scale!), or on the contrary making big noises, as they imagine and try to be "prehistoric". The ethnomusicological lesson—e.g., of ancient Pygmies in Africa or Bushmen in South Africa—tells the reverse: the more primitive (economically) the society, the more subtle its music, particularly in the main practice which is the sacred one, i.e. addressed to the Invisible. In any case, one has first to listen to the space, to its silence, to a whisper: otherwise, there is no connection with the Mystery and in this case, to the reality of the Hypogeum.

Picture 3 The Hypogeum: an architectural achievement. (Photo Daniel Cilia in A. Pace's book—see Note 1).

The question above, whether we can get some understanding of the way sound and voice were used in the Hypogeum, is obviously a hard one. However, the Hypogeum is a very resonant space; this means that the cave is in itself a musical instrument as soon as one knows how to play it.[2] From this, we can deduce some answers to the question. After a short acoustic study, we compare the results and our understanding of the Hypogeum as a whole with what is known concerning other ancient sanctuaries and particularly painted caves from the point of view of what we call the sound dimension of these sanctuaries.[3]

---

[2]This acoustic aspect of the Hypogeum has motivated the Congress in Malta; we thank here once more Linda Eneix for having organised this Congress which gave us the opportunity to study the remarkable monument.

[3]For acoustics of painted caves, see Reznikoff I., Prehistoric paintings, sound and rocks, in *Studien zur Musikarchäologie III* (Colloque international, Michaelstein, sept. 2000), Hickmann E. et al. eds., pp. 39–56, Rahden (Westf.), Allemagne, 2002.

Acoustic Study

Preliminary remark. An acoustic study of a prehistoric space is meaningful only if the space has not changed physically, and therefore acoustically. In the case of the Hypogeum, the question is not clear, since according to archaeologists and A. Pace (see Note 1), because of the burial function of the monument, in some parts, the ground was covered by dark soil while it appears now as a bare stone. Moreover, the metal staircase descending to the Middle Level, the one we studied, has certainly changed the acoustics of the centre of this level. However, because of the remarkable architectural shape and quality of stone, we can reasonably assume that the major acoustic qualities of the Hypogeum were the same, except for exact sound pitches and duration of resonance which could have been shorter. Unfortunately, we couldn't study all the rooms in the Middle level of the Hypogeum, nor the Lower Level. The study was concentrated around the centre and the so called "Oracle Room" of the Middle Level. The study was also limited in time. Therefore, this study is only a partial one; however, it gives already some insights for understanding the Hypogeum from the point of view of sound, voice and worship.

This study was carried on with the use of a male voice ranging between C_2 (65 Hz) and C_4 (262 Hz), with a normal voice, and up to C_6 or higher by producing strong overtones.[4] The pitches were controlled with a tuning fork. It is interesting to compare our results with those of Rupert Till and Paolo Debertolis, obtained by computer recording and measuring devices (see this Volume).

The main resonance of the central part (Room 17 on the Map, page 44) in the Middle Level is in A_2 (around 110 Hz), as

---

[4]N. B. In our paper mentioned in Note 3, we used the Latin numbering of octaves, while here we use the Anglo-German one.

was announced by some previous studies. This resonance is particularly strong in the "Oracle Room" (Room 18); in this room as well as in the central part and in Room 20, a secondary resonance appears in E_1 (around 82 Hz). This was expected since E_1 is the harmonic fifth of A_1 (lower octave of the main resonance). In Room 24 however, at the centre, there appears a resonance in F# (one second above E_1). The duration of the main resonance is about 4 seconds and, being in Room 17, we counted 4 echoes when using a strong voice (90 DB). We also noticed more subtle results: when moving from Room 15 to Room 17, the resonance increases at the level where the ochre painting appears (entrance of Room 17)[5]. In the Oracle Room, which is the most resonant, in the recess (or niche) on the left where ochre disks are painted, the resonance is still better; however, this recess seems not an acoustic one, otherwise it should be carved more deeply and have curved edges (so that it would be possible to introduce a part of the head). Rooms 18 and 20 are the most resonant and particularly ornate. The intensity of the main resonance is strong, the sound on A_2 is reinforced by 15-20 DB (from my own experience; it has not been measured); so that when singing 'normally' (around 70 DB) or reciting gently from the bottom of the Oracle Room 18, the whole Hypogeum resonates. Thus, considering the strength of the reinforced sound, its duration, the number of possible echoes, we can conclude that the Hypogeum has 'good' acoustics for lower sounds. However, if the resonance of the main pitches A_2 and E_1 is very impressive, the resonance is still good on frequencies up to respectively the fifth E_2 and the octave A_3 of A_2. The main range of male voices is amplified by the space. The weakness of the resonance, as compared to some parts of painted caves or to some Romanesque chapels, is that high frequencies, e.g. harmonic sounds or female high voices, are

---

[5]However, as mentioned above, the metal staircase may have modified acoustics of Room 15.

quickly dumped down, as expected because of the rather low ceiling of the rooms.

Except for the low sounds of a drum, played very softly, like a murmur, the resonance in my opinion would rather exclude musical instruments because e.g. flutes are much less amplified, while too loud a sound of an instrument would be hard or painful to hear. It is known that in ancient Greek temples as in the ancient Christian churches, instruments were excluded: no intermediate means between the praying worshipper and God.

On the Myth of 110 Hz Frequency

The question of whether 110 Hz has a particular meaning for the human being and the human brain has been discussed at the Congress. As a specialist in questions of resonance but also of sound therapy, I clearly answer by the negative. Low frequencies with their natural harmonics (overtones), if the sound is not too loud and too long, may be beneficial, e.g. the sound of the cello or of a contrabass gives a good example. But why the 110 Hz frequency? Although the anatomy and physiology of human beings are essentially the same, there are significant differences. Size, weight, heartbeat, blood pressure vary from one person to another; so does the inner resonance of the body. It is not true that there is a universal frequency on which human beings (or the whole universe!) are founded or tuned; the myth of AOUM on a fixed "cosmological" frequency that each person should find and hear for her physical or spiritual realization is denied by all professional musicians of India I have met. Each one has his or her own main tone and drone which changes with the time and years.

Concerning this frequency in the Hypogeum, in ancient times the pitch of the resonance was certainly a little different because of the level of the ground (see above) and it is ridiculous to believe that the precise 110 Hz frequency has been obtained intentionally by the Neolithic architects. There was a perpetual evolution of the rooms and levels, and the possibility to reach a precise pitch was (and is even now!) inaccessible. It is often observed that in small caves, tunnels, crypts, hypogea, there are low resonance frequencies; but this is due to the narrowness and lowness of the space. At any rate, in such spaces, if there is a resonance, necessarily there is a resonance between say F_2 (87 Hz) and C_3 (131 Hz), because resonance always appears in the harmonic fifth or octave of the main low resonance. To conclude: the 110 Hz or any other frequency could not be obtained by a special intentional achievement of the architects in this underground cave. Moreover, from the sound therapy point of view, in a given range of sounds it is not pitch that matters but intervals, particularly pure natural (not tempered) intervals and the harmonic sounds to which one listens.

A Link between Palaeolithic and Christian Era?

Because from several aspects, the Hal Saflieni Hypogeum recalls prehistoric painted caves, and at the other extremity of time, Christian chapels, it is interesting to ask the following question (the dating is intentionally approximate): *from 12,000 BC (the end of late Palaeolithic Magdalenian times) to 4,000 BC (the Hypogeum) and 2,500 BC (Egyptian temples, although the Medamoud Temple carved in the rock is dated ca. 4,000 BC), to 500 BC (Greek temples) and to 100–300 AD (the Christian painted catacombs) and, further, to 400–1,200 AD (Christian churches and chapels), is there through all these ages, a continuity?*

We leave here the question about Egyptian and Greek temples which relationship with the Maltese Megalithic architecture is obvious, but for which the acoustic dimension remains unknown. The relationship with Christian catacombs is straightforward since in some catacombs, parts were used as funeral homes and most of them were partially painted; some of them have good low acoustics, but catacombs haven't yet been studied from the acoustic point of view. The relationship to, e.g. Romanesque crypts or chapels is also clear: most of them have remarkable acoustics[6] and all the Romanesque churches were decorated with frescoes. Notwithstanding the great interest of these comparisons, we concentrate here mostly on the prehistoric period. If we choose some painted caves with good acoustics (see Note 3), what do the caves and the Hypogeum have in common?

1. In both the resonance is impressive; in the Hypogeum, particularly Rooms 17, 18, 20.
2. In painted caves, there are often niches with very remarkable acoustics and in relationship with paintings; in the Hypogeum, there is such a niche with paintings.
3. In painted caves, the most resonant parts contain more paintings; in the Hypogeum, the most resonant rooms have ochre paintings.
4. The paintings appear when the resonance increases, as in the Hypogeum, coming from Room 15 to Room 17.
5. Many objects of art, devotional or ornamental were found in the Hypogeum as well as in the caves. There is a striking similarity between the Sleeping Lady and statues of women, as obvious fertility symbols, found in caves, e.g. the famous Solutrean (ca. 18,000 BC) or early Magdalenian, "Venus" (see Picture 4) found at Lespugue (Dordogne, France) or the one from Willendorf (Austria).

---

[6]See Reznikoff I., The evidence of the use of sound resonance from Palaeolithic to Medieval times, *Acoustics, Space and Intentionality*. Identifying intention in the ancient use of acoustic space and structure. *Proceedings of the Conference held in Cambridge*, 27–29 June 2003, Lawson G. and Scarre C., eds., Cambridge (McDonald Institute for Archaeological Research, Monographs), 2006, pp. 77–84.

Picture 4 Venus of Lespugue (seen from behind). Musée de l'Homme, Paris.

Remarks. Concerning 2), we have observed above that the niche in the Oracle Room was not created for an acoustic purpose (contrary to a widespread opinion). Concerning 4), we have also observed that the metal staircase may have changed the acoustics of Room 15.

Now, appears the complementary question: what are the differences between the Hypogeum and painted caves?

1. The caves are natural, while the Hypogeum is artificial, and is an exceptional man-made architecture.
2. The painted caves were not used for funeral purposes: nowhere in such caves were human bones found.

3. Except for some signs (dots, small sticks, etc.), cave paintings are figurative and represent mostly animals, while in the Hypogeum paintings are 'abstract', representing half-spirals and discs in an arborescent way. However, there was once a painted ochre hand which has disappeared in recent years (oral communication of the former Director of Museums in Malta, Father Marius Zerafa). Concerning an interpretation of the geometric ochre paintings in the Hypogeum, we consider them in relationship with the Megalithic spirals in Western Europe (Brittany, Wales) that inspired the Irish Christian sacred Art (crosses, sacred Books) where the spirals and their interlaces represent the Spirit that comes from the cross or the Heart of Christ and blows everywhere (see Picture 5). In the Hypogeum, the geometric paintings may represent the arborescent movement, the red discs being as incarnated fruits, of the becoming souls and spirits in the visible and invisible Worlds. There is nothing similar to the spirals in painted caves.

To understand fully the significance of the Hypogeum, including the paintings and the sound, it is necessary to compare with other hypogea, for instance in continental Italy and Sardinia. Concerning the caves, it is possible to get strong conclusions from statistical results which show e.g. the relationship between acoustics and painting. But actually, as far as we know, this Hypogeum is unique for its architecture, sound and paintings.

The differences 6–8 mentioned do not affect our conclusion that the Hal Saflieni Hypogeum can be seen as a link in the long chain of sanctuaries that goes from Palaeolithic painted caves to our Romanesque crypts, chapels or churches which are, by the way, man-made architectural achievements as well. The Hypogeum, caves, temples or churches, share the same meaning in addressing the Invisible by chanting, painting, possibly dancing, and praying.[7]

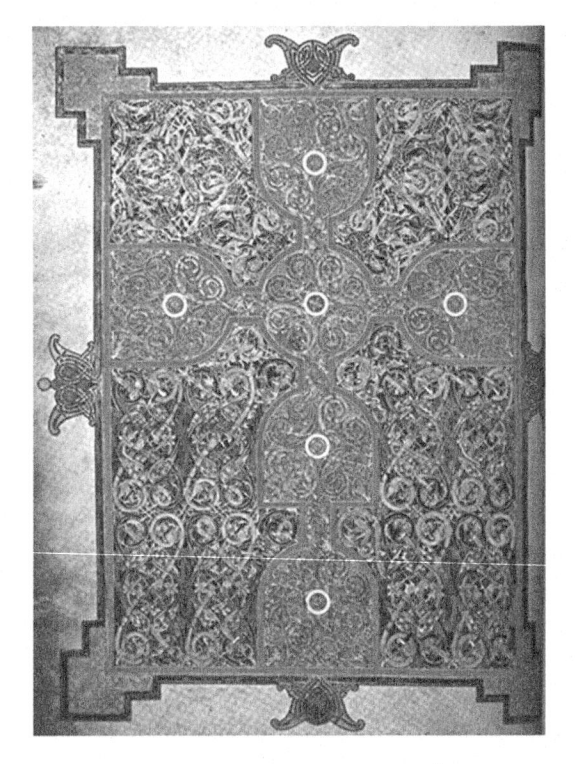

Picture 5 Book of Lindisfarne, 7th c.

This relationship between ancient sanctuaries may be based on a common, universal anthropologic need which appeared independently. But it may well be the result of slow but direct contacts since people travelled a lot by walking or

---

[7]For a possible liturgical meaning of painted caves, see Reznikoff I., L'existence de signes sonores et leurs significations dans les grottes paléolithiques. In Clottes J. (dir.), *L'art pléistocène dans le monde/Pleistocene art of the World*, Actes du Congrès IFRAO, Tarascon-sur-Ariège, Septembre 2010, Symposium "Signes, symboles, mythes et idéologie...". N° spécial de *Préhistoire, Art et Sociétés, Bulletin de la Société Préhistorique Ariège-Pyrénées*, LXV-LXVI, 2010–2011, CD: pp. 1741–1747.

floating. The whole Earth with its five continents had been settled by mankind already in the Neolithic.

That people sang laments or prayers for the dead in the Hypogeum is certain, for (a) it is a universal practice in all oral traditions we know, (b) at the same period, around 3,000 BC, we have the Sumerian or Egyptian inscriptions mentioning singing to the Invisible, particularly in relationship with death and Second Life, and finally (c) the resonance is so strong in the Hypogeum already when simply speaking, that one is forced to use it and singing becomes natural. To recite prayers is sound rewarding, especially for male voices. The deep answer of the resonance changes the personal timbre of the voice and seems to be the sound of Mother Earth or coming from the Other World: worship introduces one to deep levels of consciousness closer to the Divine.

Fear and Amazement

Torill Christine Lindstrøm and Ezra B. W. Zubrow

TORILL CHRISTINE LINDSTRØM, PhD, a psychologist and an archaeologist, is Professor at the Department of Psychosocial Science, University of Bergen, Norway.

Torill.Lindstrom@psysp.uib.no

EZRA B. W. ZUBROW, PhD, FSA, Professor of Anthropology at the University at Buffalo, USA, and University of Toronto, Canada. An archaeologist, anthropologist, and geographer, he is also Senior Research Scientist at the National Center for Geographic Analysis.

ezubrow@gmail.com

Abstract

The use of sound is surveyed for a variety of cultures and generalizations are inferred and extended by ethnographic analogy to the past. Many cultural places have acoustic properties that transform and magnify sounds. The Hypogeum in Malta is one. Sounds can create emotional effects with neuropsychological concomitants. Some sounds are fear-provoking. To subject persons to rapid successions of fear-then-relief creates so-called seesaw-of-emotions which renders those persons susceptible to comply, conform and perform. Similarly, rites of initiation also use fear as

---

Megaliths, Music, and the Mind
Linda C. Eneix
Copyright © 2024 Jenny Stanford Publishing Pte. Ltd.
ISBN 978-981-5129-25-0 (Hardcover), 978-1-003-46823-3 (eBook)
www.jennystanford.com

an ingredient, followed by a release of fear as the initiates are informed, instructed and integrated into the culture, and consequently conform to it. The parallels are striking. If the Hypogeum was used for rites of-initiation, its acoustic properties may have been instrumental in creating emotional seesaw-effects with resultant compliance and conformity.

Keywords: Sounds, seesaw-of-emotions, initiation-rites.

"Fear and amazement is a very potent combination!" said "Maximus" in the film *Gladiator*[1] referring to the spectacles of the Roman circus. Was Maximus right? And if so, does that bear any relevance for understanding places such as the Hal Saflieni Hypogeum in Malta?

If one wants to understand the past, there are only a few ways to enter it. One is through contemporary literature (history), another is through material culture (archaeology), and a third is through the biological sciences (genetics, etc.). Archaeoacoustics (the study of past sounds) is a new way to enter the past.

This paper does two things. First, it describes the role of sound in a selected set of present societies that are most relevant to the role of sound in past cultures. Second, it tries to suggest one possible explanation for the use of sound in prehistory.

The Ethnological and Ethnographic Reality

In order to understand the role of sound in past societies one suggests that modern societies that are derived from past societies will maintain some aspects of culture from their ancestors. Figure 1 shows such an evolutionary or phylogenetic tree. For example, one can trace elements of modern orchestral

---
[1]Film from 2000. Filmed in Malta.

music to earlier beginnings. So Phillip Glass (contemporary) has elements of Bach 1685–1750, Mozart 1756–1791, and Shubert 1797–1828 etc. Contemporary popular music has elements of "rock and roll", "bluegrass", "R&B", "jazz", "ragtime", etc.

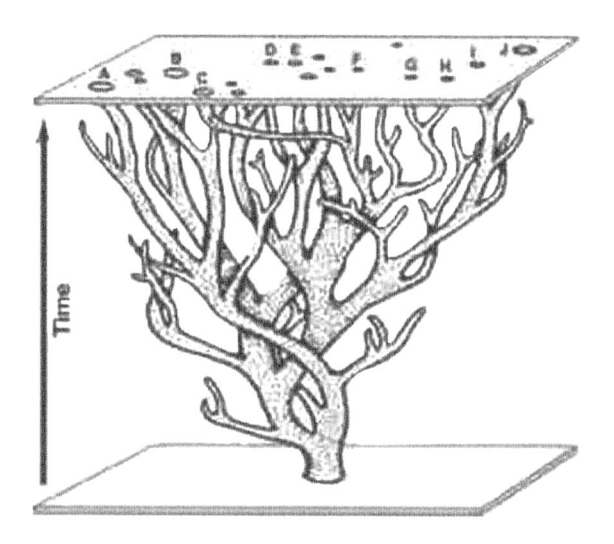

Figure 1

One may use the principles of ethnographic analogy which relates the known knowledge of present societies to unknown aspects of past societies on the basis of similarities. It is the formal interpretation of traits of past cultures based upon their similarities to analogous traits in present cultures.

For this study the Human Relations Area Files were used to analyze a large amount of comparative cultural data on sound from modern ethnographic societies (http://hraf.yale.edu/; Yale University 2014). Between the two World Wars behavioral scientists at Yale and other universities began to collect cultural materials classified at the paragraph level by subject. During World War II it was used by the Allies military intelligence to provide information about little known areas of the world where the war was taking place. After the war, a nonprofit research consortium was created to develop and distribute files of organized information related to human societies and cultures. Today it is two electronic databases filled with an increasing catalogue of cross-indexed ethnographic data, sorted and filed by geographic location and cultural characteristics. Its purpose is to foster comparative research on humans in all their variety so that explanations of human behavior would be universally valid, not culture bound.

So this study examines four domains of acoustics. They are music, speech, dance, and other types of human generated sound. Each of these domains is analyzed for characteristic sub-categories. There are eight sub-categories that are examined in relationship to the domains. They are pain, fear, emotion, startling, religion, initiation and funeral characteristics. In short, each domain is queried by asking how many ethnographic cultures use the domain for each sub-category. For example, in other words, the question is asked: "How many ethnographic societies use music in relationship to pain; how many ethnographic societies use music in relationship to fear". After completely analyzing the music domain for each of the eight sub-categories we go to the next domain: "how many ethnographic societies use speech in relationship to pain; how many ethnographic societies use speech in relationship to fear". After completely analyzing the speech domain for each of the eight sub-categories we go to the next domain. In short, we cross tabulate the four domains by the seven sub-categories. Figure 2 shows the distribution of 280 societies using the acoustic domain of music by sub-category. 150 societies use music in conjunction with pain. 85 societies use music in conjunction with funeral activity and so on.

One could ask the question does the type of subsistence make a difference. One knows for example that there are differences between hunting-gathering societies, fishing societies, pastoral societies, subsistence farming societies, and peasant agricultural

Music
OF 280 ETHNOGRAPHIC CULTURES

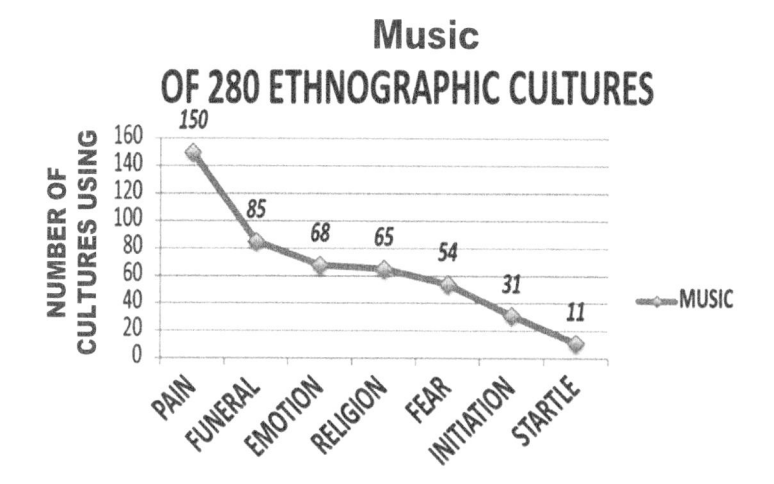

Figure 2

Non-Hunter-Gatherer Music vs. All Cultures Music
OF 280 CULTURES

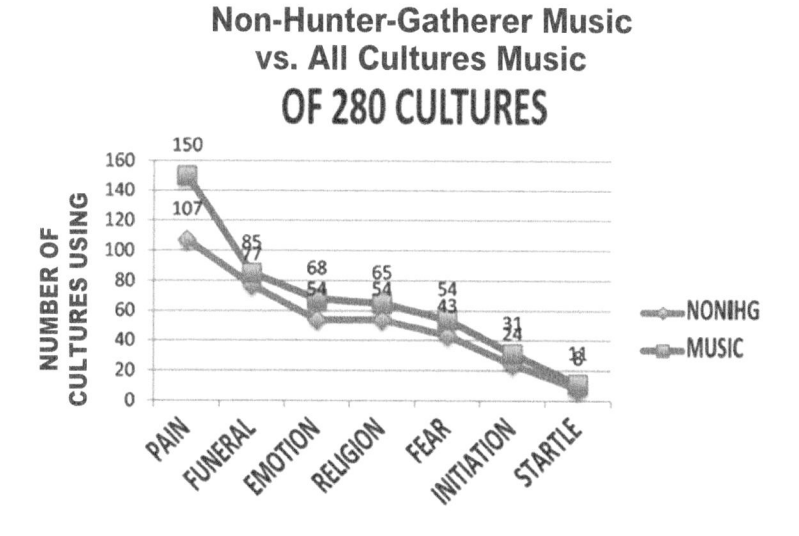

Figure 3

societies to name a few. Actually, analyzing different types of subsistence patterns does not make a difference. Figure 3 is exemplary. It shows the distribution for all "non-hunter-gatherer music" versus all societies for all seven sub-categories. The results are almost exactly the same. Of course non-hunter and gatherer societies will be less than all ethnographic societies but the basic distribution is the same and the order of the size of the sub-categories—music, funeral, emotion, religion, etc. etc., is exactly the same.

Turning to the second domain of sounds (Fig. 4), one finds a similar pattern with one significant difference. The order of the sub-categories in terms of the diminishing number of ethnographic societies is the same. In fact, the number of societies for which it is standard part of pain is even higher than in the first domain –224 to 150. However, there is a very large drop between pain and the next sub-category funeral use. In the case of music funeral use is about one half of the use with pain while in the case of sound it is slightly over a quarter.

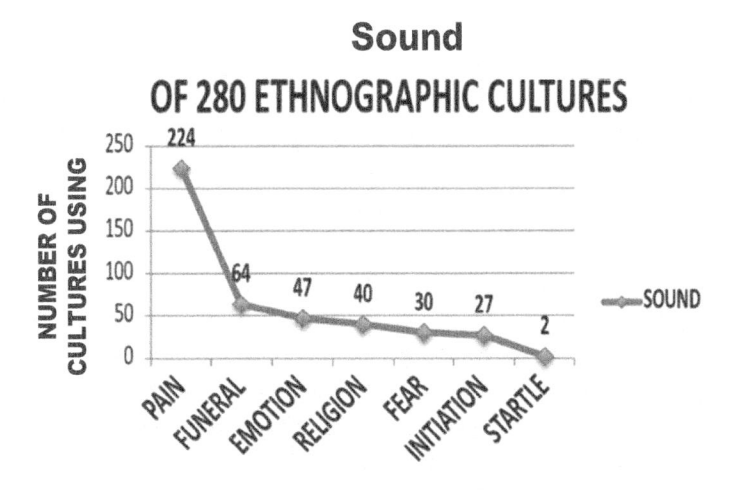

Sound
OF 280 ETHNOGRAPHIC CULTURES

Figure 4

Turning to the domain of dance (Fig. 5) one finds the same pattern but for the first and only time does one find a difference in the diminishing order of the sub-categories. Use of dance in funeral activities fits between fear (135 societies) and initiation (89) societies.

Dance
OF 280 CULTURES USING

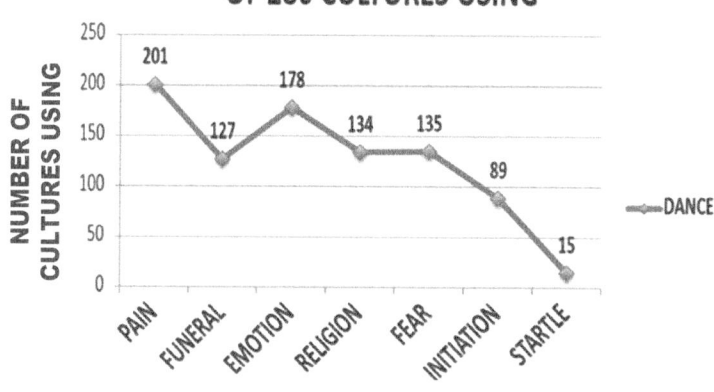

[Chart showing number of cultures using dance: PAIN 201, FUNERAL 127, EMOTION 178, RELIGION 134, FEAR 135, INITIATION 89, STARTLE 15]

Figure 5

Turning to speech one sees the same order again. The relatively small numbers is caused by the rather formal definition of speech.

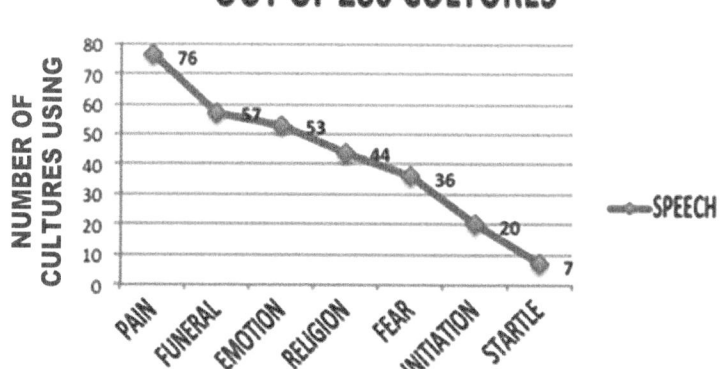

Speech
OUT OF 280 CULTURES

[Chart showing number of cultures using speech: PAIN 76, FUNERAL 57, EMOTION 53, RELIGION 44, FEAR 36, INITIATION 20, STARTLE 7]

Figure 6

Figure 7 combines the four domains. It simply reinforces what has been previously noted that the patterns across the domains are the same.

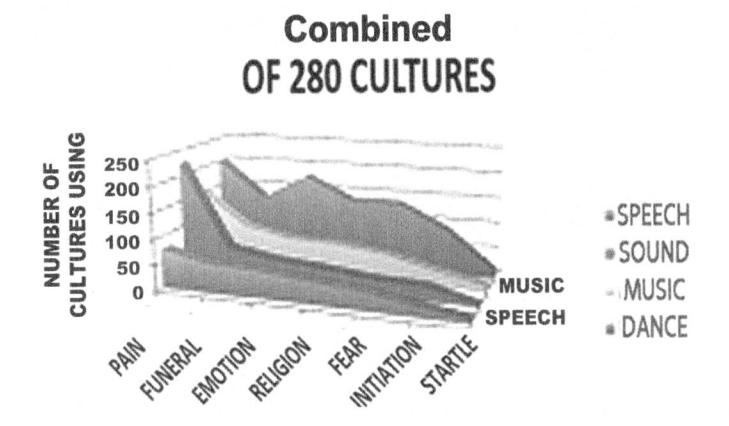

Combined
OF 280 CULTURES

Figure 7

In addition to the similarity of patterns, one may note some other results. They are that pain consistently is the most frequent category across the domains averaging approximately 58% of all cultures. Using sound for startling people is consistently the least frequent averaging 12.5%. The use in funerals are twice more frequent than in initiation-rites and ceremonies (30% to 15%). Perhaps, the most unusual domain is dance with its relatively high emotional context.

One need not make too much of the above results. There are a variety of potential problems. Some of these are more significant than others. There were analytical rules used for the cross tabulating of the number of cultures in each sub-category in each domain. Here is an illustration. If in a particular society there are several domains used in single instance of a sub-category that sub-category is counted for each domain. For example if in a particular initiation ceremony of a particular culture there is dance, music, sound, and speech, it was counted as an instance for each domain. So in the Na'ii'ees, the Western Apache Woman's Puberty Ceremony in Phase 1 the pubescent girl dances, in Phase 2 singing begins, in Phase 3 massaging takes place, in Phase 4 running occurs, in Phase 5 running in all directions, in Phase 6 gifts poured over her head, in Phase 7 blessing speeches, in Phase 8, she leaves a woman. So the Na'ii'ees would be counted as one for each domain. One might want to argue this is a type of multiple counting. It is only one ceremony but it is being counted in each domain.

Similarly, in a particular society's funeral ceremony there is a large amount of speech and only a few minutes of music. The rule that is used in this study is that both count equally and they are each counted as an instance of each domain. However, one could try to have a weighted system. However, there also are problems with that. What should be the criteria for the weighting? Should it be comparative time? Something that is short may be more important than something that is long. For example, a few minutes of musical prayer may be far more critical than a eulogy. There are other weaknesses as well. One could argue that the definitions are not equivalent. For example, pain in one society may not be equivalent to pain in another.

However, even so the patterns of similarity across such a large number of ethnographic societies should give us pause and does lead credence to the common ancestry of sound usage in prehistoric society. The authenticity of common ancestry and the diminishing order of the sub-categories are probably as far as one wishes to go and not push the conclusions much further. Yet it is quite evident that "sounds" in various forms and formats have been, and are, essential to human culture. In all places and through all times people have used sounds of their own creation in connection with important rites and essential emotional experiences.

Still, one needs to raise the question: *Why* did people produce and use sounds? This paper suggests *one* possible use of sounds, for a particular purpose.

Inside Stone Sanctuaries

Sanctuaries in caves are as old as humanity itself. Venues in caves for rituals and cultic practices are found worldwide (Reznikoff 2004/2005). In Mediterranean contexts caves were places for rites through all prehistoric periods from the Paleolithic (Campbell 1976, Mavridis et al., 2013).

Caves could be changed and decorated for such purposes. Among those, the Hypogeum of Hal-Saflieni in Malta, is perhaps the most unique and outstanding. From around 4000/3300 BC, over a period of at least 1500 years, it was extended with numerous rooms and galleries going wider and deeper into the ground, and developed into a large and complex subterranean sanctuary. Skeletons of approximately 7000 individuals were found in the Hypogeum, so evidently it was used as a communal burial place or "knuckle-house" for a long time. However, that does not preclude that it was also used for other cultic and ritual purposes, before, during, or after its function as a necropolis.[2] What ritual that might have been, we can only carefully and tentatively suggest on the basis of comparative anthropological and archaeological information.

The Hyper-Acoustics of the Hypogeum

The Hypogeum has many perplexing properties, one of which is the acoustics. They have caused much speculation regarding the ritual uses of the Hypogeum, of ritual uses of the acoustics in themselves, and even to speculations about whether the

---

[2] Also Christian churches are used for very different ceremonies.

Hypogeum was built out (or rather: carved out) in order to create or develop its particular acoustics. The acoustics have already been explored to some extent, and some findings will be outlined here. In the early works of Devereux et al. (1995) and Jahn et al. (1995), and in the more recent works of Devereux (2009) and Cook et al. (2008) it is concluded that many of the ancient solid stone structures, (called "temples"), in Malta, and in particular the Hypogeum, have resonances registered at the frequency of approximately, at 110–111 Hz. Similar deep pitch frequencies have been registered also in prehistoric megalithic monuments elsewhere (Watson and Keating 1999). Likewise, many caves containing rock art are documented to have strong resonant acoustics (Reznikoff 2004/2005). These acoustic properties appear to be found particularly in places where rock art is located, and seem to be surprisingly unrelated to the surface-properties of the stone.

Yet, according to the findings of the explorative investigation performed by the Achaeoacoustic Conference in Malta on February 21th, 2014, there are also several other resonances than the 110–111 Hz to be found in the Hypogeum, with frequencies as different as 68–70 Hz (male voice) and 114 Hz (male voice, drum) (Archaeoacoustics 2014).

Still, the 110–111 Hz frequency has, perhaps erroneously, been regarded as *the* frequency of the Hypogeum, and therefore investigations regarding this particular frequency have been done. Experiments done by Cook et al. (2008) recorded that when people were exposed to sounds at the pitch of 110 Hz, it produced certain alterations in the brain's electrical activity. The activity patterns over the prefrontal cortex, measured by EEG, were affected and changed. The language centre was to some extent de-activated, and there was a temporary shifting from left-hemispheric to right-hemispheric dominance. Several other frequencies were tested, but this finding was reported to be particular for 110 Hz. And, it was explicitly commented in the report that the right side of the prefrontal cortex is related to emotional processing, meaning that such acoustics made people particularly aware of emotions.

In the information on the web (www.disclose.tv/forum/the-underground-chambers-of-the-hypogeum-t...) about the Hypogeum it is claimed that: "A word spoken in this room (the so-called "Oracular Room") is magnified a hundredfold and is audible throughout the entire structure." And: "The effect upon the credulous can be imagined when the oracle spoke and the words came thundering forth through the dark and mysterious place with terrifying impressiveness." This refers to the indisputable fascinating acoustic properties of the Hypogeum, but whether oracular activities took place there, is simply an assumption. But, apparently, according to these quotes above, *not necessarily a particular pitch* was essential, but also a *high volume* of sounds was needed for producing the auditory effects and experiences of the Hypogeum—whatever these effects and experiences might have been. We will suggest a few.

Effects, Experiences, Emotions

Words spoken would be audible "through the entire structure". Yet, Cook et al. (2008) concluded that their findings of changes in brain-activity were compatible with a relative de-activation of the language centres which are located in the left hemisphere of the brain; and contrastingly, there was a shift towards right-hemispheric dominance. Further, that a prefrontal activity related to emotional processing took place. These findings do of course not rule out the possible importance of spoken words: speech, prayers, chanting, or singing, as important sounds made in the Hypogeum. But they *could* mean that the *emotional* effects of vocalizations, less than the meaning of words, might have been more at the core of the experiences. And indeed, not only voices, but also other sounds caused by movements, footsteps, whistling, animals, and music-instruments may have sounded and resounded in the Hypogeum.

During the experiments performed in the Hypogeum (Archaeoacoustics 2014), it was our impression that all sounds, but in particular dark sounds were magnified. People respond with orientation reflexes, even startle reflexes, to sounds (and other stimuli) that are: sudden and strong (Sokolov 1960, Davis et al. 1982, Lang et al. 1990). One could call the response: "*amazement*". We would suggest that *dark* sounds are particularly likely to trigger such responses. Sudden and strong sounds elicit fear even in new-borne babies. This is a reflexive response. Human hearing, through our phylogenetic development, is adapted to be particularly aware of sounds signalling danger (Schaal 2012). Many danger-signalling sounds are dark, sudden and strong, (such as the sounds of predators, avalanches, earthquakes, thunderstorms, and aggressive human vocalizations), (Thompson 2002). Danger evokes *fear*.—Given this, we then have the combination of "*fear and amazement*" as effected by sounds.

Is fear located in in a particular part of the brain? Yes and no. The amygdala is essential for fear, and various areas of the brain show different activation-patterns in relation to different emotions. But, psycho-neurological research, using methods such as EEG (electroencephalography), fMRI (functional magnetic resonance imaging), and PET (positron emission tomography), all being various forms of brain-scanning, tell us that despite that there are particular areas in the brain which are more active in connection with certain emotions the whole brain is largely involved in all emotions, and there are individual differences (Phan et al. 2002, Kassam et al. 2013). Still, with regard to lateral dominance patterns, it is claimed that people who tend to have a left-hemispheric dominance are more set to experience positive emotions such as: joy, happiness, amusement, and positive social emotions; whereas people with a right-hemispheric dominance have a bend towards negative emotions such as: depression, worry, and anxiety (Schwartz et al. 1979, Ahern and Schwartz 1985, Alfano and Cimino 2005).

The right hemisphere has a predominant role in mediating attention, perception of emotional stimuli, and of interpreting visual and *auditory* stimuli that are relevant for emotional processing. The right hemisphere is also important in the regulation of the autonomic and endocrinological activation that takes place with emotional arousal, (Lane et al. 1999). The association between negative emotions such as anxiety, fear, worry, and a preponderance of right-hemispheric activity, means that the observed shift from left to right-sided prefrontal cortical dominance in subjects exposed to sounds at 110 Hz (one of the resonances of the Hypogeum), does support our suggestion that "fear and amazement" were likely emotional experiences to be had in the Hypogeum.

Given these premises, it is possible to imagine this scenario taking place in the Hypogeum in ancient times: First: if there was a sudden shift from silence to sound, it would elicit an immediate surprise-reaction, a startle-response ("amazement"), with a subsequent sudden orienting reflex. Startle-responses and orienting reflexes are in themselves intensely attention focusing, and have a particular short latency-time when the stimulus is auditory (Lang et al. 1990). Then, after this startle and amazement, given that the sounds made and emitted were around 110–114 Hz, (or simply any dark pitches), and of a strong volume, such sounds were likely to produce a right-hemispheric dominance in the audience. That would produce a set towards emotional attention and facilitate a situation where, in particular *fear*, but also other negative emotions could be elicited and experienced. In other words: It is not unlikely that ancient people inside the Hypogeum could have experienced the combination of "fear and amazement".

Yet, if that really was so, the question still remain: Is "fear and amazement" "a potent combination"?—Psychologically speaking?—Does it have any other effect on the human mind apart from being a strange and otherworldly experience, a "kick", or "a thrill?"

Seesaw-of-Emotions

We believe so. In a series of experiments (both field- and laboratory-experiments) Dolinski, Nawrat and collaborators (1998, 2002, 2007, 2013/14, 2013, 2013) demonstrated that persons who were exposed to a situation with stimuli that elicited an emotional response, and then, quickly thereafter, were exposed to stimuli that triggered a different emotional response, which rendered the first response irrelevant, became "mindless" and bewildered. And, shockingly, those persons became subsequently surprisingly susceptible to influences, and would tend to comply with any suggestions, requests, and whatever was demanded of them. They became easy to influence, convince, and command. One could perhaps call the effect a light brain-washing. In the first, and most well-known, experiments the sequence of emotions was that of "fear-then-relief". People were scared, and then shortly thereafter were informed that there was no danger, resulting in an emotion of relief and even joy. Later, experiments have shown that also combinations of other emotions can give the same results. The pivot point is the rapid shifts and quick requests of compliance shortly after. Dolinski called this phenomenon "The Seesaw of Emotions", (referring to a children's playground-instrument that makes children go up and down in rapid successions). The explanation given for the seesaw-effect is that several emotional programs of opposite contents are activated in such rapid succession, that a person's brain (and the person) becomes emotionally and cognitively dissonant, bewildered and mindless. Persons such "softened" will therefore tend to respond positively to anything that serves at a clue about what to do in the situation. Social requests are strong cues, so when such "bewildered" people were asked to give money, help somebody, buy something, etc., they tended to be surprisingly willing to agree and comply.

The effect of seesaw-of-emotions has been documented with various combinations and shifts of emotional content.—But, since *fear* is such an unpleasant, intense and spellbinding emotion, with profound physiological concomitants, we believe that the fear-then-relief-combination is a particularly powerful ("potent") one. The consequence of seesaw: making people susceptible to accept and obey messages and commands, and to conform, can be very useful indeed for those who want to influence people or exert power. We therefore suggest this effect as a possible and plausible motive for creating seesaw-of-emotions and *using* them for manipulative purposes.

Rites-of-Initiation

Rites-of-initiation (rites-de-passage) (Pausanias (in Jost 1985), Van Gennep 1960[1909], Campbell 1976, 2008[1949], Eliade 1976, Ustinova 2013) are found in almost all human cultures, cross-culturally, trans-historically and pre-historically. They could take place in caves in prehistoric times (Campbell 1976). Despite variations and different types, they generally imply these elements: initiatory ordeals, separation, seclusion, darkness, blind-folding, pain, fear, symbolic-death-experiences; then revelation and/or instruction of truths or of mystic wisdom, insight, enlightenment, epiphany of the Deity, blind-folding, darkness and fear dispersed, concluding with a return to the ordinary society and/or introduction into a secret society.

Psychological laboratory experiments have shown that initiation-rites can produce *cognitive dissonance* with resultant feelings of strong group *attraction* among initiates after the experience (Festinger 1962). Initiations can also produce *conformity*, cohesion and compliance among members of the society (Harmon-Jones 2002). It strikes us that rites-of-initiation have many perplexing similarities with the sequence of events and psychological effects of the seesaw-of-emotions phenomenon.

Could *sounds* be used for fear-induction in both processes? —We think so.

Could the Hypogeum be used for initiation-rites?—We think that could be possible.

Resounding the Arguments

However, words of caution are warranted. The ethnographical data showed that sound was seldom associated with fear. Yet, intentional, ceremonial affliction of fear and pain may be underreported or masked in other phrases. To instill fear using sound is, although perhaps an infrequent phenomenon, still well-known from music worldwide and from modern film soundtracks (Ryan Drake 2012).

We are, of course, fully aware of the fact that the use of psychological manipulations in order to increase peoples' susceptibility was not scientifically studied in the prehistoric past. Nevertheless, we believe that the effects of manipulative strategies may have been discovered intuitively and used strategically. With regard to the possibility of sounds having been used in the Hypogeum in order to elicit fear, we admit that this possibility is highly speculative. There is absolutely no evidence that the Hypogeum was a place where sounds, vocal, instrumental, or others, were deliberately emitted for any particular purposes. The Hypogeum certainly has fascinating acoustic properties. But we find no reason to believe that the Hypogeum was created and built out in order to enhance, develop, and use the acoustics for cultic, ritual purposes. However, that being said, we regard it as almost inevitable that people in the Neolithic past in Malta discovered the acoustic effects of the Hypogeum, and experienced these as extraordinary strange, perhaps even as weird and "otherworldly". As a consequence, since the acoustics were already there, people may have found the acoustics useful in cultic ceremonies and rites. If so, the uses of the acoustic could be regarded as a "secondary product" of the Hypogeum's properties.

The only *documented* use of the Hypogeum was as a "graveyard", housing for the dead. Yet this did not preclude other uses of the premises. Quite the contrary. The presence of the community's dead could be effectful in combination with the fear-provoking potential of strong sounds, sounds that could be "magnified a hundredfold" within the Hypogeum. That combination renders itself as a very useful implement to create the components of fear and symbolic death that are characteristic of initiation-rites—seesaw-of-emotions—*"fear and amazement"*.

Conclusion

Admittedly, our suggestion is a chain of "ifs". Still, we find the parallels between the ingredients and effects of seesaw-of-emotions and of rites-de-passage as very striking: Seesaw-combination of: fear, then relief, and ensuing social compliance; and initiation-rites with: fear-inducing symbolic death, and finally, reassuring acceptance *into* the social order, along with compliance, conformity and acceptance *of* the social order.

Still, initiation-rites may never have happened in the Hypogeum. Yet it does offer great locations for initiation or other rites. The dead bodies that were stored there certainly made a perfect "set" for experiencing a symbolic death and a descent into the netherworld. The darkness and acoustic effects of the place could heighten the experience of fear, and finally, the ensuing return to daylight and the relief of having "made it", would indeed be perfect contexts for creating seesaw-of-emotions in initiates. Increased social compliance would be the likely result. It is important to note that this might be the effect on simply anybody entering the Hypogeum and getting safely out into the light of the upper world again. Initiation-rites are a possible, but not a necessary prerequisite.

But, why?—Is there reason to believe that social compliance and submissiveness was treasured in the Maltese Neolithic context?—We believe so. The Neolithic itself was characterized by cultures focused on a new invention: collaborative agriculture. And, the many astounding temples in Malta, and the Hypogeum itself, bespeak of enormous collective collaborations over extended periods of time. For these large-scale projects of agriculture and building social cohesion and compliance was absolutely needed and necessary.—To this end, *"fear and amazement"* could be *"a very potent combination"*.—"Maximus" was right.

References

Ahern, GL, and Schwartz, GE. 1985. Differential lateralization for positive and negative emotion in the human brain: EEG spectral analysis. *Neuropsychologia,* 23, 6, 745–755.

Alfano, KM, and Cimino, CR. 2005. Alteration of expected hemispheric asymmetries: Valence and arousal effects in neuropsychological models of emotion. *Brain Cognition,* 66, 213–220.

Archaeoacoustics 2014. Preliminary report. http://www.otsf.otg/conference2014.htm.

Campbell, J. 2008 [1949]. *The Hero with a Thousand Faces.* Novato: New World Library.

Campbell, J. 1976. *The Masks of God: Primitive Mythology*. New York: Penguin.

Cavalcanti, A. 1900. Sound in films. In Weis & Belton (eds) *Film Sound: Theory and Practice*, pp. 98–111. New York: Columbia University Press.

Cook, IA, et al. 2008. Ancient architectural acoustic resonance patterns and regional brain activity. *Stud Time*, 1, 10, 95–104.

Davidson, RJ (ed) 2000. *Anxiety, Depression, and Emotion*. New York: Oxford University Press.

Davis, M, et al. 1982. A primary acoustic startle circuit: lesion and stimulation studies. *J Neurosci*, 2, 791–805.

Davis, BP, and Knowles, ES. 1999. A disrupt-then-reframe technique of social influence. *J Pers Soc Psychol*, 76, 192–199.

Davis, BP, and O'Donohue, WT. 2004. The road to perdition: "Extreme influence" tactics in the interrogation room. In: O'Donohue & Levensky (eds) *Handbook of Forensic Psychology*, pp. 897–996. New York, Elsevier.

Devereux, P. 2009. A ceiling painting in the Hal-Saf lieni hypogeum as acoustically-related imagery: a preliminary note. *Time and Mind*, 2, 225–232.

Devereux, P, et al. 1995. Acoustical properties of ancient ceremonial sites. *J Sci Explor*, 9, 438.

Dolinska, B, and Dolinski, D. 2013. Peur-puis-soulagement, légitimité d'une petite contribution et comportement charitable. *Rev eur Psychol Appl*, 64 (2014), 29–34.

Dolinski, D. 2013. On the seesaw. When the danger is over. *J Russ East Eur Psychol*, 50, 3, 65–79.

Dolinski, D, and Szczucka, K. 2013. Emotional disrupt-then-reframe technique of social influence. *J Appl Soc Psychol*, 43, 2013–20141.

Dolinski, D, et al. 2002. Fear-then-relief, mindlessness, and cognitive deficits. *Eur J Soc Psychol*, 32, 435–447.

Dolinski, D, and Nawrat, R. 1998. "Fear-then-relief" procedure for producing compliance: Beware when the danger is over. *J Exp Soc Psychol*, 34, 27–50.

Eliade, M. 1975 [1958]. *Rites and Symbols of Initiation*. New York: Harper & Row.

Erikson, M. 1964. The confusion technique in hypnosis. *Am J Clin Hypn*, 6, 183–207.

Festinger, L, et al. 1956. *When Prophecy Fails*. Minneapolis: University of Minnesota Press.

Festinger, L. 1962. Cognitive dissonance. *Sci Am*, 207, 93–107.

Harmon-Jones, E. 2002. A cognitive dissonance theory perspective on persuasion. In: Dillard and Pfau (eds) *The Persuasion Handbook: Developments in Theory and Practice*. pp. 99–116. Thousand Oaks: Sage.

Human Relations Area Files. http://hraf.yale.edu/; Yale University. 2014.

Jahn, RG, et al. 1995. Acoustical resonances of assorted ancient structures. *Technical Report PEAR 95002*. Princeton: Princeton University.

Jost, M. 1985. Sanctuaires et Cultes d'Arcadie. Paris: Librairie Philosophique J. Vrin.

Kassam, KS, et al. 2013. Identifying emotions on the basis of neural activation. *PLOS ONE*, 8, e66032.

Lane, RD, et al. 1997. Neuroanatomical correlates of pleasant and unpleasant emotion. *Neuropsychologia*, 35, 1437–1444.

Lane, RD, et al. 1999. Common effects of emotional valence, arousal and attention on neural activation during visual processing of pictures. *Neuropsychologia*, 37, 989–997.

Lang, PJ, et al. 1990. Emotion, attention, and the startle reflex. Mendeley. com. Retrieved 2011-10-01.

Mavridis, F, et. al. 2013. Anonymous cave of Schisto at Keratsini, Attika: A preliminary report on a diachronic cave occupation from the Pleistocene/Holocene transition to the Byzantine times. In: Mavridis and Jensen (eds) *Stable places and changing perceptions: Cave archaeology in Greece. BAR International Series 2558*. Oxford: Archaeopress.

Nawrat, R, and Dolinski, D. 2007. "Seesaw of emotions" and compliance: beyond the fear-then-relief rule. *J Soc Psychol*, 147, 556–71.

Pausanias. VIII, 23, 1.

Phan, KL, et al. 2002. Functional neuroanatomy of emotion: A meta-analysis of emotion activation studies in PET and fMRI. *Neuroimage*, 16, 331–348.

Reznikoff, I. 2004/2005. On primitive elements of musical meaning. *J Music and Meaning.* http://www.musicandmeaning.net/issues/showArticle.php?artID=3.2.

Ryan Drake, J. 2012. The importance of sound design and its affect on perception. Master of Art Thesis. State University of New York at Fredonia, Fredonia, New York.

Schaal, H-J. 2012. Das rettende Organ. Prähistorische Funktionen des Ohrs prägen unser Musikhören bis heute. *Neue Z Musik,* 173, 5, 34–36.

Schwartz, GE, et al. 1979. Lateralized facial muscle response to positive and negative emotional stimuli. *Psychophysiology,* 16, 561–571.

Sokolov, EN. 1960. Neuronal models and the orienting reflex. In: Brazier (ed) *The Central Nervous System and Behavior.* pp. 187–276. New York: Josiah Macy.

Thompson, B. 2002. Evoking terror in film scores. *M/C: A Journal of Media and Culture,* 5. http://www.media-culture.org.au/0203/evoking.php.

Ustinova, Y. 2013. To live in joy and die with hope: experiential aspects of ancient Greek mystery rites. *B Inst Class Stud,* University of London, 56, 105–123.

Van Gennep, A. 1960 [1909]. *The Rites of Passage.* Chicago: University of Chicago Press.

Watson, A, and Keating, D. 1999. Architecture and sound: an acoustic analysis of megalithic monuments in prehistoric Britain. *Antiquity,* 73, 325–336. www.disclose.tv/forum/the-underground-chambers-of-the-hypogeum-t.

Sound, Cognition, and Social Control

Vincent C. Paladino

VINCENT C. PALADINO is a certified audio engineer, a cognitive anthropologist and an independent researcher.

Abstract

The question of intentionality is central to interpreting our findings in Archaeoacoustics, the question of whether or not ancient peoples, literate and pre-literate, possessed detailed knowledge of acoustic phenomena tasks us to search for answers that help contextualize the subject. Additionally, the issue of social control has been addressed within Archaeoacoustics, with questions arising from the notion that knowledge of acoustic phenomena might add to the prestige and power of an elite. The intention of this paper is to contribute to the establishment of a theoretical context within which our observations and experimental results can be framed, positing the notion that anthrophonic sound is a mediator of social control and its intentional use is the result of adaptive cognitive processes. The use of sound in the construction and maintenance of social systems allows for the effective manipulation of the environment. Social control is defined here as the necessary establishment of structure and form which assists in the establishment of more

---

Megaliths, Music, and the Mind
Linda C. Eneix
Copyright © 2024 Jenny Stanford Publishing Pte. Ltd.
ISBN 978-981-5129-25-0 (Hardcover), 978-1-003-46823-3 (eBook)
www.jennystanford.com

predictable outcomes. The use of sound in a conscious fashion is understood here as an inevitability, driven by the cognitive and physical needs of humans.

Humans use their voices and bodies to transmit and receive information, and a great deal of it is in the form of sound. This information has explicit and implicit components, some conscious and others unconscious. We inquire of our territory and inform our environment, interrogate our surroundings and fill space through our use of sound.

Sounds of all kinds emanate from humans, anthrophonic sounds like grunts, screams, groans, speech and music. There are work sounds, play sounds and the sound of people sleeping. Our rituals and rites need sounds to empower them, and the use of sound even empowers the application of silence. A quote famously attributed to Miles Davis posits:

"It's not the notes you play, it's the notes you don't play."

Thus all sacraments and traditions, sacred and secular, acknowledge sound's importance, even those performed in silence. There is the persistent image of the one-eyed witchdoctor, but he's never deaf. "A noteworthy fact is that a deaf ritual specialist is an anomaly whereas a blind one is common enough" (Jackson 1968:296). Sound is crucially important, because it is a means of contact and control.

The purpose for which humans make sound is unique. We do more than signal some present condition, we indicate the past and future as well. We pattern it into languages spoken, written and musical that broadcast feelings and inspire the formation of ideas. These patterns transmit knowledge, accumulated facts and insights that are passed on through generations, driving technological and social developments that would not be possible without such collected information.

Sound resonates with great power in our lives. It has been a vitally important component of our development, its power within the developmental dynamic well demonstrated. Throughout our evolutionary process we have had an ongoing exchange with and through sound. This evokes questions like whether pre-literate people rendered paintings in resonant cave spaces intentionally and whether their leaders or their shaman utilized acoustic characteristics to control and guide people's decisions and actions.

The position taken here is that sound has been used with knowledge and intention throughout human history, with the degree of knowledge and technical skill varying through time and territory. Furthermore, people use sound as a tool of social construction and maintenance, a set of actions that may arise from the functional characteristics of human consciousness. Significantly, the idea of intentional action may have to take on a dimension of inevitability or a deterministic quality given the structure and function of the human mind/body/brain.

We use and have used sound for social control because it is one of the vibrational components of an energetic exchange within our environment to which we contribute directly using our bodies. We live in a complex relationship of energy exchange and reorganization of matter that manifests through many unknown and hidden causes, driving common events, our survival depending on an ability to construct models of the world and make accurate predictions. In effect, imposing order that allows us to survive. Controlling the structure and function of social groups is a method of ensuring predictable outcomes based on defined and established relationships. Social control is an effort toward survival, an adaptive behavior that includes within its myriad entanglements the emergence of a powerful elite and a definition of people as a resource to be utilized in the pursuit of other scarce resources. The minimization of prediction error entails the construction of a controlled environment which requires restriction and control of individual and group behavior, a goal served well using language and music. This process can be examined in a structured way using theoretical constructs formulated to explain and predict the functions of human cognition and its action in the environment.

Free Energy

When utilizing the information based paradigm of human cognition as formulated by Karl Friston (Friston, 2007), communication through sound is understood as a tool used to minimize free-energy. Free-energy is an overall measure of surprise experienced when observations diverge from one's mental model of the environment. This model is expressed by equation below (Friston, 2007):

$$F = -\ln p(\tilde{s}|m) + D(q(\vartheta|\mu) \,||\, p(\vartheta/\tilde{s}))$$

Free-Energy = Surprise + Cross Entropy

F	Free-energy
$-\ln p(\tilde{s}|m)$	The upper bound of surprise associated with receiving a sensory input: $\tilde{s}$
m	Model of the world
ϑ	Unknown causes, or unknown quantity that caused the sensory state
μ	The internal states of the brain
$q(\vartheta|\mu)$	Recognition density
$p(\vartheta|\tilde{s})$	True distribution of unknown causes: ϑ
$D(q(\vartheta|\mu)||p(\vartheta|\tilde{s}))$	The divergence between the Recognition Density and the True Density

Cross entropy is the divergence between the recognition density and the true distribution of the causes of a sensory state (Collell and Fauquet, 2015).

The minimization of prediction error is an adaptive characteristic of consciousness, a functional quality that plays a crucial role in the organism's survival within its environment. The construction of an accurate mental model of the world, along with active reduction of danger, is an ongoing process. The environment is continually sampled through the senses, the model adjusted, and/or the sensory input is changed through action within the world (Friston, 2010). Such actions include the creation of social systems that control people and their environment.

Using tools to transform the material world and transduce surrounding energies to meet their needs is a power unique to humankind. These processes begin with sound, which is used to create social cooperation and generate effective group activity. Myths, religions, science, laws, taboos and customs are all social constructs which make life a more understandable and safer experience. Human culture is a network of systems that serve the purpose of creating stability and predictability. Vocal sounds, music and language transmit the knowledge and emotional drive to create the needed social and physical structures that sustain us. This relationship between sound, cognition and social structures inspires the development of intriguing areas of inquiry and research that can be pursued by Archaeoacousticians. Let's consider some of these.

Internal to External Structures

Through an information exchange between the internal and external environments, there is mutual construction and modeling taking place between mind and environment. The idea that social constructs and the hierarchical structures in the brain are creating each other is a fascinating study in environmental adaptation and neural representation in the exosomatic and endoneural worlds. Endoneurally, the very shape of neurons is a clue to the nature and structure of the universe. Action at a distance, such as gravitational influence between objects in space, is modeled by the long slender nerve body of a nerve cell, which connects axon to dendrite (Friston, 2017, 10:33). The characteristics of the universe in which the brain exists are cast within its structures. Another area of consideration is how social structures model the brain's communication network, moving information between people as the brain does between neurons. Here, brain functions are modeled in the surrounding environment when the system is driven by intentionality in the form of sound and symbols.

Sound Transmits Brain Activity

Sound is a medium that couples our brain activity to the external world. Music and spoken language carry vibrational information from the brain to the outer environment, intended to order that environment in a manner that allows us to continue the feedback process of experimentation, construction, learning and growth. When we speak, talk, sing and play instruments, we are reducing free energy (surprise) by acting on the environment to change its states to a more ordered, human-friendly and predictable form, which then changes our sensory input. We change the world around us to get from it the feedback we need and consequently expect, on a consistent basis.

Consider the temporal structure of periodicity. Oscillations in the brain are periodic, and the periodic waves that emanate from us as sound can be used to create social cohesion. Periodicity is experienced as regularity resulting in predictability. Where periodic phenomena occur, a conscious agent can predict occurrences more accurately. People communicate to share feelings and thoughts, to form alliances and groups that act as force multipliers. They make music, a rhythmic set of periodic impulses, to express themselves and the experiences they share with their group. This builds cohesion and fosters effective cooperation to build society and its structures. In this way, the regularity of periodicity is used within the system to reduce uncertainty. What beautiful symmetry!

It is well known that periodic waves can interfere with each other destructively, cancelling each other out. We know that social interactions can be harmonious or disharmonious. Evolution itself may be characterized as a progression driven by constructive and destructive interference. Those changes-mutations-that are harmonious result in constructive interference propagating in a wave-like fashion through the biosphere, and those that are dissonant result in destructive interference and are cancelled.

Beliefs, Myths, Social Control

In caves, reverberant sounds might be interpreted as emanating from within a rock, and myth supporting that would reduce uncertainty and fear of the unknown. A model is thus provided that reduces surprise. The same would be true of resonant spaces within buildings of religious worship, such as cathedrals. The glory, magnificence and wonder of the building, and by extension the universe, can be attributed to the workings of the worshiper's deity. Those who worship in such a place would be receiving confirmation of their beliefs with every encounter. Their model of the world is reinforced each time they experience the sounds and sights in the cathedral. The energy input the system (cathedral + people) needs to continue reassuring worshipers that their world model is correct is largely in the form of sound. Sermons, songs and the anthrophonic sounds of people in groups supply auditory and physical sensations that facilitate bonding, which reinforces shared beliefs. In this way and in this environment, uncertainty is reduced.

Discussion

Did people of ages past use sound knowingly, and as a means of social control? Utilizing these models of cognition and brain activity, it appears that the answer is a likely yes. Initially, they may have done so without conscious awareness, driven by the need to reduce free energy through manipulation of their mental models and their sensory input data. However, as their awareness grew, their conscious power of manipulation would have grown. During the initial condition, intention may be defined as actions taken as the result of internal cognitive processes, separate from the consciousness occupied with everyday tasks. During later conditions, as awareness developed progressively, this intention became the actions taken as the result of observation and calculation within the conscious mind of everyday life. Social control and the intentional use of acoustic effects emerge from a system that seeks to verify its beliefs, operating internally at frequencies that can be externalized as sound. The sounds we make are transmissions of culture, and the study of this process can lead us to greater understanding of the interrelationship of our brain, mind, social structures and the universe.

References

1. Friston, K. J., and Stephan K. E., Free-energy and the brain. *Synthese*, 2007, **159**(3): 417–458. *PMC*. Web. 9 Jan. 2018.
2. Friston K., *Nat. Rev. Neurosci.* February 2010, **11**(2), 127–38. doi: 10.1038/nrn2787. Epub 2010 Jan 13. Review.

3. Collell, G., and Fauquet, J. Brain activity and cognition: a connection from thermodynamics and information theory. *Front. Psychol.*, 2015, **6**, 818. doi: 10.3389/fpsyg.2015.0081.
4. Atasoy, S, Donnelly, I., and Pearson, J. Human brain networks function in connectome-specific harmonic waves. *Nat. Commun.*, 2016, **7**, 10340. *PMC*. Web. 9 Jan. 2018.
5. Jackson, A. Sound and ritual. *Man* New Series, Jun 1968, **3**(2), 293–299.
6. Friston, K., Free energy principle [video]. June 16, 2017. YouTube. https://www.youtube.com/watch?v=NIu_dJGyIQI.

Voicing Cave: Experience and Metaphor, from Archaeoacoustics to Voice Therapy

Irén Lovász

DR. IRÉN LOVÁSZ, PhD, is an associate professor in the Institute of Arts Studies and General Humanities at KRE University in Budapest, Hungary. Her research has included ethno-musicology, anthropology of religion and music, sacred communication. She is also a professional singer, applying traditional singing in voice therapy.

www.lovasziren.hu

Abstract

The author shares some experiences and experiments regarding the effect of pure human *voice* under the Earth in the natural acoustics of ancient *cave*s, undercrofts, and in rotundas, basilicas and also in Turkish thermal baths with cupola structure. In some cases the strong resonance of the sacred architecture was used as a natural acoustic *instrument* itself. An unusual sound behaviour can be experienced in the 12 Apostoles' Rotunda of *Bény* (Slovakia), 12th Century, with *12 mysterious vaulted niches* in it. The author also introduces a method, *cave* as a metaphor used in contemporary *voice- therapy*, including basic notions and images of archetypical symbolisation.

Megaliths, Music, and the Mind
Linda C. Eneix
Copyright © 2024 Jenny Stanford Publishing Pte. Ltd.
ISBN 978-981-5129-25-0 (Hardcover), 978-1-003-46823-3 (eBook)
www.jennystanford.com

Keywords: Cave, voice, acoustics.

Introduction

The acoustic dimension of the *human voice* in sacred and spiritual communication and in voice-therapy is one of my main fields of interest. In spite of the fact that "the magic practice of singing with echos is forgotten,"[1] as a professional singer, I myself prefer singing in acoustic spaces like caves, Romenescque and Gothic churches, and rocky mountains in nature. I really try to sing in a resonance, that is fully using the acoustic character, as Professor Reznikoff underlines, "we need to sing in just intonation, i.e. in the natural pure intervals of resonance."

As a scholar studying the anthropology of religion and music, I highly appreciate that recent studies in the multidisciplinary field of archaeoacoustics opened a new door on understanding the use of *voice* with focus on ancient use of sound in sacred and contemplative spaces.[2] I personally welcome the fact that researchers are now taking note of unusual sound behaviour in the world's monumental places. I do my best to advance the scholarly acknowledgement of this important *paradigmatic shift*[3] also in my home country, Hungary.[4]

---

[1] Reznikoff, Iegor: The evidence of the use of sound resonance from palaeolithic to medieval times, Lawson, G., and Scarre, C. (eds.), *Archaeoacoustics*, University of Cambridge, Cambridge 2006, Chapter 8. 83.

[2] See: http://www.otsf.org/background-reading.html (2015.11.20.)

[3] See: Zubrow, Ezra, B.V.: The silence of sound. A prologue. In Eneix, L. C. (ed.), *Archaeoacoustics: The Archaeology of Sound: Proceedings from the 2014 Conference in Malta*, OTSF, Florida, 2014. 7–9.

[4] Lovász, Irén: Barlangtól a kupoláig - Szakrális terek akusztikai hatásairól az arhaeoakusztikától a hangterápián át a kortárs előadóművészetig. (*From Cave to Cupola—Acoustic effects of Sacred Spaces. From archaeoacoustics and soundtherapy to comtemporary performing arts*). In Sepsi, Enikő; Lovász, Irén; Kiss, Gabriella; Faludy, Judit (eds.), *Proceedings of the Conference on Art and Religion*, KRE - L'Harmattan 2016.

I have to admit that I was impressed by the studies of Professor Igor Reznikoff on the sound resonance from the Palaeolithic to Medieval times. We met in Prague at a conference: on *Human voice as a tool of sacred and spiritual communication*, organised by Soundscope in November 2014.[5] After I read his articles, I found the material of the Malta conference organised there a few months earlier, February 2014, by the OTS Foundation.[6] I got so enthusiastic about the scientific *synchronicity*, since the Hypogeum in Malta belongs also to *my exceptional personal experiences* and a great revelation too! I experienced the mystical acoustics there in 2010, and since that time I have been keeping the picture of the Hypogeum before my eyes to see it all the time. These were the main reasons that brought me to the 2nd Multidisciplinary Conference of Archaeoacoustics to Istanbul.

In this paper first of all I would like to share my own experiences and experiments—both as a singer and a scholar in anthropology of religion and music—using human voice in acoustic sacred places. I will then describe how such experiences and experiments have been used in practical applications in therapeutical contexts.

Acoustic Experiences

My first experience regarding the unusual behaviour of the human voice and its psychic and physical effects in special acoustics happened in the beginning of the year 2001. I felt like singing in one of the old Turkish baths in Budapest, in the *Saint Lucas thermal bath*.[7] It happened in the main chamber with a *cupola* of a perfect half globe shape reminding me of

---

[5]http://www.hlasohled.cz/eng/other/conference.html (January 11, 2016).
[6]Eneix, Linda. C. (ed.): *Archaeoacoustics: The Archaeology of Sound: Proceedings from the 2014 Conference in Malta,* OTSF, Florida, 2014.
[7]Budapest is famous of its healing thermal water and thermal baths. Some of them were built in the 16th century during the 150-year Turkish period of the country, in the style of the traditional Turkish baths.

Nordic *iglu* or the Eur-Asian *jurta*. I was standing in the middle of the space. There was a small, round whole opened to the sky above my head, and the bright sunshine came into the pool from there in direct rays. I felt like I was standing as the extension of Axis Mundi in the water which is the most universal element of purification. I kept my mouth under the water. The sound of my humming was transmitted strongly by the water, and reverberating strangely but nicely by the ancient walls of the kupola in concentrated strong echos in the mystical, steamy surrounding of the bath. The natural elements combined and worked together strengthening each other in high concentration. It turned out soon that not only I, but people sitting around me in the bath by the wall, definitely felt the positive effect of this energy in our body and soul in that collective experience, which reminds me of the archaic medieval sacred practice of, *incubatio*. People were not aware of the source of the mystical sound. They were not sure whether the source of the sound was in the water, or in the air, or in the old walls, or came from above by way of the sun rays. I repeated this experiment several times, while finally it became clear that it was me who made that sound with the inevitable help of the special acoustics of the cupola and the natural elements: Water, Earth, Fire and Air. After a while people could join me, some of them started to sing with me. But this led me to the next step of my scholarly path.

My second main acoustic experience regarding the power of sound in acoustic spaces happened in the same year (2001) in the thousands of years old ancient caves on *Elephant Island* in India, near Mumbai. When I started to sing, I immediately found the pitch, the resonance of which has the strongest and the most direct effect on my body and soul. This resonance coming back from the walls of the cave for my voice reached the most sensitive point of my body around the Heart-chakra. I did not care whether it was the 110 Hz frequency or not. But I definitely had the immediate experience of the supernatural, spiritual power of the human voice in acoustic, sacred space. I had the idea that that might have been the effective (and that is why) the most common pitch of producing for instance the sound of OM in the Buddhist traditions for thousands of years. Or this could have been the frequency of reciting effective mantras, prayers or of the ritual verbal formulas and sacred songs too.

The *Elephanta Caves* are a network of sculpted caves located on Elephant Island, or *Gharapuri* (literally "the city of caves") in Mumbai Harbour, 10 kilometres (6.2 mi) to the east of the city of Mumbai. The island, located on an arm of the Arabian Sea, consists of two groups of caves—the first is a large group of five Hindu caves, the second, a smaller group of two Buddhist caves.[8] The rock cut architecture of the caves has been dated to between the 5th and 8th centuries. The caves are hewn from solid basalt rock. All the caves were also originally painted in the past, but now only traces of the paintings remain. The main cave (the Great Cave) was designated as a UNESCO World Heritage Site in 1987 to preserve the artwork.

Thirdly I would like to draw the attention to some of the medieval sacred places in Central Europe in the Carpathian basin, where unusual sound behaviour can be experienced. One of them is the 12 Apostoles' Rotunda of *Bény* (Slovakia), from the 13th Century, with 12 *mysterious vaulted niches* in it. Each of the 12 niches strengthens different resonances, which gives a special sound to the songs here. All of them are different in size, and very probably tuned on purpose. If you sit in the first one and put your ear to the wall, you hear the lowest sound, and at the last one you can hear the highest sound. There is probably a definite awareness of resonance and the conscious application of artificial niches into the architectural

---

[8]http://whc.unesco.org/archive/advisory_body_evaluation/244.pdf. (January 11, 2016).

construction in order to improve sound quality thus serving spiritual purposes since, niches, recesses or alcoves were used as natural resonators[9] in medieval architectural construction.

On my field trip in 2013. I conducted some interviews with the local people there. According to them the acoustics in the building changed a bit in that church, unfortunately, due to the recent renovation. A 40-year-old priest who spent all his childhood in the village told me that one of his favourite activities as a small boy was sitting and singing in the niches while testing to see in which one of them his feet could reach the floor.

To discover the secret of the mystical acoustics of this medieval rotunda and the niches in it, I am going to organise a multidisciplinary scholarly team, including experts in architecture, acoustics, early music, anthropology of voice and arts in the near future.

Another special medieval acoustic sacred space, the undercroft of the Basilica in Pécs inspired our concert: *Sacred voice and sacred space* with my singer, researcher colleague Zoltán Mizsei. The history of the Cathedral of Pécs goes back to the period of the Roman Empire. The base walls of the undercroft were built at the end of the 4th century and they were extended in a westerly direction in the 8th–9th centuries. The original church became an undercroft in the period of King St. Stephen (11th century) as a sanctuary was built above it. I myself could feel the strong energy, and the exceptional acoustics with my whole body there. The protective power and the surroundings of that marvellous underground sacred space offered such a strong experience for us that this served as the inspiration and the basis of our Sacred Voice cd.[10]

---

[9]Reznikoff, Iegor: The evidence of the use of sound resonance from Palaeolithic to Medieval times. In Lawson, G., and Scarre, C. (eds.), *Archaeoacoustics*, University of Cambridge, Cambridge, 2006, 80.

[10]Lovász, Irén: Sacred Voice (Healing Voices 1.) SVCD01. Siren Voices, Budapest, 2006.

The Healing Voices cd Series

During the past several years I have been experiencing both as a singer and as a cultural anthropologist researching folk religiosity and sacred communication to what extent the *human voice* is able to serve as an aid in life-and-death situations and in other defining events that shape and turn one's fate all over the world. In 2006 I started to publish a four-part cd series entitled *Healing Voices*. The first is Sacred Voice, the second is Inner Voice, and these will be followed by Female Voice and Healing Voice. These four are linked by the shared idea that the human voice may very well have healing powers for this is, indeed, one of the most important ancient and universal functions of singing.

The *Sacred Voice* cd contains mostly archaic Hungarian folksongs which have survived as a living oral tradition and are still being sung in the most archaic regions of the country. The order of the songs represents the festive circle of the year. The accompaniments are free interpretations of the so-called "Burdon technique", ranging from *"overtone singing"* to the repetitive *ostinato*. This is known throughout numerous cultures the world over. Its role is to help the listener submerge him- or herself deeper into the musical conjuring of the *eternal*. Additionally, we have chosen instrumental and vocal improvisation, also inspired by the ambience of the *enchanting acoustics of the sacred architecture*, where the recording sessions took place. It was the *undercroft* of the Saint Stephan Basilica in Budapest.

The natural acoustics of that sacred space under the Earth can be felt on the recording with its long echos, reverbs on the sound. The strong resonance of the sacred architecture was used as a natural acoustic *instrument* itself.[11]

---

[11] Lovász, Irén: Csendességben/In silence. https://www.youtube.com/watch?v=JXhSBhCcQbo (January 12, 2016.) The natural acoustics of the Basilica can be experienced not only in the undercroft but also in the main

Inner Voice is the second chapter of "Healing Voices".[12] It is a meditative inner journey meant to awaken the primeval powers slumbering at the very depth of the "Self". It brings to the surface the energies of the ancient elements of earth, water, fire and air through the natural means of music and harmony. We can only reach the very depth of our being through the help of cultivating silence completely, a turning inward in meditation. Tapping into the primeval powers slumbering in the depths of our "Self" occurs when we use the most ancient of string instruments, that is, the *human voice,* and special instruments that awaken the four archetypical elements. We represent the Earth through the Australian aboriginal wooden instrument, the *didgeridoo,* Water through the Celtic *harp,* Fire through the Hungarian *violin,* and Air through the Japanese *shakuhatchi.* I have collected powerful ancient Hungarian folksong texts and melodies representing each of the elements.

To my mind the sequencing of the four elements draws the map of the development of a given personality. From the point of view of this article, only the first two elements, the Earth and the Water are the relevant ones:

Earth: This is the place before one's birth, conjuring up the mother *womb*'s life-giving *cave*. The earth is where roots live. This is the source of our stability, providing the possibility of life itself. Its colour is brown.

Water: This represents *amniotic fluids*. The water-mill helps the grinding away of personal sorrows, their surfacing to consciousness, and their eventual elimination. Its colour is blue. Water is the most reassuring, most helpful element that loosens up everything, as it reminds one of the amniotic fluids. Water is the original, universal medium for purification.

---

space of the bulding under the cupola. Life performance of the Sacred Voice cd, 2007, Budapest. St. Stephan Basilica.

[12]Lovász, Irén: Inner voice (Healing Voices 2) SVCD02. Siren voices, Budapest, 2007.

Cave-Womb-Soundtherapy

Since the concept includes the archetypical representation of *cave* as a universal symbol of femininity- by recalling notions of Mother Earth, mother's womb, it can well be applied in psychotherapy of female disorders. There are a few examples of the practical application of my *Inner Voice* cd in different therapeutic contexts also in clinical circumstances. The most remarkable has been done by Dr. Bea Pászthy, Associate Professor, Head of Child and Adolescent Psychiatry, Department of Paediatrics, *Semmelweis University, Budapest*:

> "We use your *Inner Voice* recording to treat depression and also in the healing of teen age girls suffering from Anorexia Nervosa. During an intensive week-long therapy we apply the Inner Voice cd for evoking the symbols and archetypical features of the Elements.
>
> Each day we focus on one of them, and try to relate the given personality to each of the Elements, balancing in this way the harmony of body and soul, child and family, past and present, nature and woman, since *feminity as a whole* finds itself in great crisis in all these cases. *Female power* and values should be recalled and strengthened through the *archetypal feminine attitudes*. I am convinced we found a right way using your Inner Voice cd for these healing purposes."[13]

The *physiological* basis of the healing power of voice is that we are sensitive to different noises, sounds, voices already in our mother's *womb*. The embryo can produce physical reactions not only for the relaxing but also for the disturbing sounds or musical effects. That is why it is so important to be aware of

---

[13] Lovász, Irén: Power of sustainable heritage through the healing voice of traditional singing. In Hoppál, M. (ed.), *Sustainable Heritage*, European Folklore Institute, Budapest, 2010. 227–238.

the type or quality of music a mother listens or sings during her pregnancy. The first sign of life is a sound: the *heartbeat*, which is rhythm and resonance. The first obvious sign of life of a newborn is a sound, a *human voice*. That is a crying voice of a new human being. The first and most effective tool of relaxing, calming down a baby, besides hugging and rocking, is the rhythm and the energy of the mother's singing voice. The *Lullaby* is the traditional tool for calming, relaxing a baby. I myself sang personally improvised lullabies every day for each of my children until they reached the age of ten.

The efficacy of voice therapy is based on the fact that our *deepest consciousness* is structured by *sound*. By deep levels of consciousness we mean the very first levels acquired in early childhood and even before birth. With the exception of the sense of sight, the means of perception, particularly the auditory system of the child in its mother's womb are already formed at the sixth month of pregnancy.[14] The richest and most structured perception is the perception of sound and noise: aural perception by way of the ears—in a liquid milieu the vibrations are very well perceived—but also, as we have seen, perception by way of the entire body. The body perception of sound by sensing vibrations belongs of course to the deep levels of consciousness and we are usually unaware of it, for instance when we speak or even sing. It is important to notice that the passive perception of sounds preserves continuity with the period in the womb.[15] Professor Reznikoff also combines very sensitively the notion of *cave* to the notion of *womb* on a very archaic level of human consciousness: "It is as

---

[14] See: Reznikoff, Iegor: On primitive elements of musical meaning, *JMM: The Journal of Music and Meaning* 3, Fall 2004/Winter 2005 [http://www.musicandmeaning.net/issues/showArticle.php?artID=3.2], sec. 2.9.2.4 (January 12, 2016).

[15] Reznikoff, Iegor: On primitive elements of musical meaning, *JMM: The Journal of Music and Meaning* 3, Fall 2004/Winter 2005 [http://www.musicandmeaning.net/issues/showArticle.php?artID=3.2], sec. 2.4 (January 12, 2016).

if the same cave, the same sanctuary, the same cathedral was rediscovered or reconstructed ever since our first and deepest sanctuary: in the womb of our Mother—women, Earth or God."[16]

Cave as a Metaphor

That is the basis also of the method I myself have been using in voice therapy. I am going to introduce the method through which I use *cave* as a metaphor in contemporary *voice-therapy*, including the notions of *Mother Earth—Mother Womb* as it appears in the archetypical symbolisation.

I started an experimental Therapeutic Singing Circle in 2008 which meets every week by the Danube in the centre of Budapest. This is an experimental workshop where I call people, mostly women to sing with me in a group every week.

In each meeting we do ritual welcoming and boundary-marking voice training, which is a *voice scale* within the body. I also call it inner body-massage with one's own voice. The scale or *ladder metaphor* helps to imagine the self being bound between Earth and Heaven, as the person, pronouncing the sound as the Axis mundi. At the same time this vocal exercise helps us to experience our own voice as an instrument of communication of the inner self to the supernatural.[17]

---

[16]Reznikoff, Iegor: The evidence of the use of sound resonance from Palaeolithic to Medieval times. In Lawson, G., and Scarre, C. (eds.), *Archaeocoustics*, Cambridge University, Cambridge, 2006, Chapter 8. 83.

[17]This inner body massage conceptually was inspired by a retreat I experienced in Hungary, led by two voice therapists from UK, Cambridge: Felicity *Cook* and Browen *Rees* in 2007. I am very grateful for both of them for the inspiration. Especially for giving me the unexpected inspiration for creating my own MANTRA, which is on my Inner Voice cd as the last track: https://www.youtube.com/watch?v=KpOi8V5M7lU&list=PL6F3D48D9748501D5 (January 12, 2016),

> Mother Earth under my feet,
> Father Sky above my head,
> I, myself, the axis,
> I, myself, the axis.
>
> My roots hold me with their strength,
> My roots hold me down secure
> Anchored in the strong wind,
> Anchored in the strong wind....[18]

In the weekly singing circle we have been doing the voicing scale, elaborated into my own experimental method. Besides the scale we also work with the *cave as a metaphor*: that is with imagination of our resonant parts of our body like our mouth as a *cave* washed by water from inside. Our voice is the water. It has to catch all the micro chambers of the resonant cave, in order to achieve or cover the greatest resonant surface of our body. We pronounce only vowels like: *a, á, o, ú, ó,* or *ó, a, á, o, ú,* or *e, é, ö, ü, ő*. We change the resonant surface of our mouth-cave by changing back-front-mid vowels, and the position of our mouth that is the entrance of the *cave* rounded, unrounded, closed, middle or open.[19]

It is easy to understand and to experience the *movement of sound in the body*. First, the higher sounds of the voice vibrate in the higher parts of the body (throat and lower part of the head) while lower ones vibrate in the lower part of the body (chest and back). It is a very simple and convincing experience to put one's ear on the upper middle part of the back of a person and listen to the sound this person produces, the sound moving, let us say, in the range of a fifth: it goes up and down

---

[18] Lovász, Irén: Mantra (detail). In Inner Voice cd, booklet linernotes SCVD02 Sirenvoices 2007.

[19] We made a short demonstaration of this method during the second *Conference on Archaeoacoustics* in Istanbul together with the participant collegues. Singing together in the auditorium was a very special experience for most of us.

along the spine and the back of the singer. And hence it is not purely conventional when we say that sounds are *high* or *low*: on the contrary, it is a reality based on the *body perception of sound*. Moreover, because of its movement the sound structures the spine and therefore the body; this structuring can be shown to be precise. This was about the movement of sound related to pitch. But there is a second movement, independent of pitch: the movement of the vibrations of different vowels and consonants in the body.[20]

It is also important to link the *voice-ladder* exercise and the *ladder–axis metaphor* we apply in the voice therapy to the function of *spine* in the *body perception* and also in the *body creation* of sound. Of the two perceptions of sound we have, either in the body or auditory, the first one is actually the most important. Indeed, those born deaf do perceive sound vibrations in their body, and they can learn to speak and even to sing when trained properly in this corporal perception, without any auditory apparatus; whereas, not to perceive any sound vibration in the body is a sign of mental backwardness. This proves that our perception of sound is founded on the corporal perception of vibrations.[21]

Focusing on the role of the body in singing, Imke McMurtrie gives examples of different cultures and traditions from Sufism to the Georgian polyphonic singing. In India the Voice is called the *"The Queen of all instruments"*, because the singer has to *tune* his or her voice and lend the whole body to the music...If the body is tuned like this, it is ready to

---

[20]Reznikoff, Iegor: On primitive elements of musical meaning, *JMM: The Journal of Music and Meaning* 3, Fall 2004/Winter 2005 [http://www.musicandmeaning.net/issues/showArticle.php?artID=3.2], sec. 2.3 (January 11, 2016.)

[21]Reznikoff, Iegor: The evidence of use... *Proceedings of the Konference Hlasohled*, Prague 2014.71; Iegor Reznikoff, On primitive elements of musical meaning, *JMM: The Journal of Music and Meaning* 3, Fall 2004/Winter 2005 [http://www.musicandmeaning.net/issues/showArticle.php?artID=3.2], sec. 2.4 (January 12, 2016.)

open, receive and send sound vibrations in an optimal state (eutonus) without blocking, dampening or distorting the vibration of the bodily tissue (100 billion cells in the human body and their state can hinder or increase the vibrational quality of the singer). Our bones are the most crystallised form of connective tissue. Its chemical formula is very similar to wood (instruments). The finest and highest vibration can be reflected in the densest cellular structure.[22]

Professor Reznikoff also argues that *the first consciousness of space is given through sound*. The child doesn't see but hears the voice of the mother high or low in her body...and the sounds or noises in various locations coming from internal or external surroundings. This sense of space is important for the child to position itself in the right way, head down, in preparation for the moment of birth. It has been shown that children whose mothers sing are in general better positioned for this major event.... The relationship of the consciousness of space with sound perception is demonstrated also by the fact that people who are born deaf have difficulties in comprehending space. To conclude, in early levels of consciousness, space and sound, and therefore movement and change of sounds, are deeply related, and hence, in our consciousness music and rhythm are inseparable from movement and space.[23]

Regarding the relation between space and movement of sound, Reznikoff makes a remarkable suggestion: In a prehistoric cave, one of the most impressive experiences is to discover the cave, walking in complete or almost complete darkness, and all while making sounds (preferably vocal ones) and to listen to the answer of the cave. In order to figure out

---

[22]McMurtrie, Imke: Singing is a way of prayer. in *Proceedings of the Voicescope Conference on: The Voice as an Instrument of Communication with the Transcendent*, Prague, October 8–9, 2014, 57–59.

[23]Reznikoff, I: On primitive elements of musical meaning, *JMM: The Journal of Music and Meaning* 3, Fall 2004/Winter 2005 [http://www.musicandmeaning.net/issues/showArticle.php?artID=3.2], sec.2.9 (January 12, 2016).

where the sounds come from—from far away or from nearby—and whether there is somewhere a strong resonance or not: all this in order to ascertain the direction in which one may proceed further on. ...This way of moving around in darkness demonstrates the main importance of sound in discovering space and in proceeding through it; to be sure, it reminds one of the first perception of space the child has in the world of the mother's womb.[24]

Silence, Sound and Resonant Chambers in Voice Therapy

Not only space and sound, but also silence and darkness are deeply related. Why, then, are people so afraid of silence and remaining in darkness? This keeps them from hearing their own inner voice, although we could gradually find our way back to quietude and silence.

From the point of view of communication, the power of the acoustic message is stronger, at least much more direct and straight if there is no disturbing noise in the communication channel during the transmission. In other words: silence is the best environment for effective communication of sound.

I call good music the extension of silence. If we cultivate outer and inner silence then we will find our own inner voice. What we perceive as, "good music" is the result of the harmony of spiritual and physical vibrations. Physical sound waves reverberate on the, strings of our souls. The healthy person's vibrating surface is in harmony. When we become ill the balance is disturbed and we have to rebalance it so that we can heal. This is how singing has a therapeutic role. While singing with the help of our own sound waves we use our

---

[24]Reznikoff, Iegor: On primitive elements of musical meaning, *JMM: The Journal of Music and Meaning* 3, Fall 2004/Winter 2005 [http://www.musicandmeaning.net/issues/showArticle.php?artID=3.2], sec.2.5 (January 12, 2016).

bodies as a resonating cavern or chamber or *cave*. Under these circumstances we can harmonize our own bodily and spiritual vibrations.

What is it that keeps us from rediscovering our own voice, resounding out from our own silence, from our own soul? To find the answer we might make further excavations in the field I tried to describe above that provide us a few important insights regarding the relations between the archaic human notions of *cave, voice, acoustics, resonance, sacred spaces*, and contemporary ideas on *voice perception, and voice therapy* also. I believe that if we can examine all aspects of ancient traditions of the use of sound and space to effect consciousness, then we can see how all this directly relates—not only to the design of contemplative architecture—but also to other potential applications for today[25,26], like for instance contemporary voice- and sound therapy, in order to preserve, advance and in some cases to reconstruct the mental and physical wellness of humanity.

---

[25] See: http://www.ancient-origins.net/events/second-international-multi-disciplinary-conference-archaeology-sound-002683#ixzz3xJb9bWQe (January 15, 2016).

[26] Lovász, Irén, Female voice as a tool of sacred communication and healing in the Hungarian tradition. Konference Hlasohled 2014 Hlas Jako Nástroj Komunikace S Tím, Co Člověka Přesahuje Praha 8-9 Listopadu 2014. http://www.hlasohled.cz/uploads/pdf/SBORNIKY%20-%20KONFERENCE/sbornik_konference%202014.pdf (January 11, 2016).

Initiation: Inside the Great Pyramid

Steven Halpern

What follows is a very personal account of Grammy®-nominated recording artist, composer, pioneering sound healer, author, and researcher Steven Halpern. Mr. Halpern is considered one of the founding fathers of new-age music. Although not presented in a conference environment, Mr. Halpern has generously agreed to include his experience and impressions in this volume.

StevenHalpernMusic.com

I came into this lifetime hardwired for transcendence, and have been fascinated by the effects of deep reverb and echo throughout my life. My first experience of the consciousness-altering power of deep reverb in large, acoustically charged chambers evoked in me an altered state of consciousness and past-life memories, though I didn't have that vocabulary at the time to describe or understand what I had experienced. Thus began my intuitive quest to seek out ways to access that state again.

My first experience of the consciousness-altering power of reverb to transform a single "note" into a three-dimensional

---

Megaliths, Music, and the Mind
Linda C. Eneix
Copyright © 2024 Jenny Stanford Publishing Pte. Ltd.
ISBN 978-981-5129-25-0 (Hardcover), 978-1-003-46823-3 (eBook)
www.jennystanford.com

tone happened quite by accident. In 1962, my first year in high school, I snuck into the empty auditorium and played one long note on my trumpet. The note seemed to hang in the air forever. It had a richness I had never heard before. I recall being aware of an unusual feeling in my head. I liked the feeling. It felt like "home." (Years later, I learned that I had just experienced a balancing and synchronization of the two hemispheres of my brain.) I sat there in the silence and the darkness, wondering what had just happened. And how could I experience that again.

In the 1880s, German scientist Hermann von Helmholtz coined the term "psycho-acoustics" to describe the effects of sound and music on our physical, psychological, and spiritual well-being. But it wasn't until the 1970s that scientists like Dr. Gerald Oster began publishing. In Oster's research paper (*Auditory Beats in the Brain*) the deeper realities behind the ancient use of tone and rhythm began to explain the underlying processes.

It would be several decades before the concepts of brainwave synchronization and mindfulness would enter mainstream and holistic medical consciousness. Today, we can simulate and stimulate higher consciousness with sonic technology and music. That's the area I've focused on all of my professional life.

Itzhak Bentov, a brilliant scientist, inventor, and author of the seminal book on entrainment, *Stalking the Wild Pendulum*, invented specialized equipment that could measure subtle responses that no other scientist of his day (1970s) could measure. Even today, few can match some of his discoveries. I was honored to have him help support my graduate dissertation research project involving Brainwave Biofeedback and Kirlian Photography and the effects of classical vs my own original meditative compositions. Bentov describes how the phenomenon of rhythm entrainment and brainwave entrainment activate a phase coupling with the fundamental electro-magnetic field of the earth (the Schumann resonance).

I believe this to be the key to a theoretical and historical understanding of the connection between megalithic structures and music: a net effect that would bring participants into a state of higher coherence that they would describe as getting into closer communion with the divine.

In 1981, I had a chance to investigate this for myself when I was invited to participate in a tour of Egypt to give workshops and headline a concert in front of the Sphinx. I looked forward to feeling what it would be like to meditate inside the King's Chamber of the Great Pyramid.

I promised Mahmoud, the chief guard, that if he allowed me and my recording engineer to have an hour or two by ourselves, he could be part of the recording that I wanted to make. (Of course, there were some other aspects of the negotiations that contributed to gaining his permission.)

I bought the state-of-the-art portable analog tape recorder and hired my studio engineer to professionally record not just the sounds, but the acoustic pressure and "silence" inside the King's chamber. Thus, I had a once-in-a-lifetime opportunity to be alone inside the Pyramid, chanting tones, and sacred names. I explored the amazing harmonics and resonances. We did acoustical readings of standing waves as well, generated by my voice and flute. This recording is now available as *Initiation: Inside the Great Pyramid*. (Listening with headphones is the closest you can come to experiencing the magical resonance and entrainment of the King's Chamber without physically being there.)

As soon as I began to chant, I began to feel like I was levitating. I could feel the energy field around my head shift into a higher order of resonance and synchronization.

On the final day of our tour, I had an unexpected last opportunity to get do additional recording. This time, I climbed into the sarcophagus, lay down, and began chanting tones. Within seconds, I felt like I was in a small boat, rocking on the ocean. Again, I began to feel like I was levitating. There was more that I wanted to explore. But just then, the spell was broken. Apparently a group of Japanese tourists had convinced the chief guard to let them enter, even though I was supposed to be alone for an entire hour. Their noisy voices meant it was time for me to move on. Today I do my best to help listeners achieve hemispheric balance and higher coherence in the alpha and theta brainwave state through my recordings.

Visit Youtube.com/StevenHalpernMusic to see videos of cymatic imaging by John Stuart Reid of Steven's recording "Initiation: Inside the Great Pyramid."

Index

Absolute Unitary Being (AUB) 153
acoustic aesthetics 178
acoustic analysis 184
acoustic anomalies 11
acoustic archaeology 157
acoustic awareness 57
acoustic character 242
acoustic characteristics 162, 233
acoustic dimension 202, 242
acoustic effectiveness 91
acoustic effects 31, 43, 186, 226–227, 238, 242
acoustic engineering 35
acoustic environment 14, 81
acoustic evaluation 12
acoustic experiences 243, 245
acoustic experiments 15
acoustic phenomena 190, 231
acoustic pressure 259
acoustic properties 11, 43, 160, 209–210, 219–220, 226
acoustic qualities 199
acoustic research 15
acoustic resonance 39
acoustic standing waves 127
acoustical readings 259
acoustical resonance 37
acoustically engineered performance environments 43
acoustics 11, 39, 111, 131, 157, 195, 198–200, 203–205, 212, 218–220, 226, 242–247, 256
 architectural 177
 good 96, 203
 mystical 243, 246
 natural 241, 247
acoustification 96
ADHD, *see* attention deficit hyperactivity disorder
after death world 196
agriculture 4, 57–58, 61, 70, 159
 collaborative 227
all cultures music 213
amniotic fluids 248
ancestors 14, 31, 35, 41, 51, 64–65, 68–70, 104, 127, 210
ancient ritual structures 39
animals 52, 68, 70, 74, 88, 103, 125, 141, 205, 221
anterior cingulate cortex 188
anthropology, cultural 5
archaeoacoustic evaluations 33, 78, 149
archaeoacoustic research 15, 40
archaeoacoustic studies 10, 12, 150
archaeoacoustic techniques 107
archaeoacousticians 32, 235
archaeoacoustics 5, 10, 12–14, 83, 159–160, 162, 164–166, 175, 210, 219, 221, 231, 241–256
 impact on prehistory 49
archaeological research 4, 12, 106, 203
archaeologists 9–10, 13, 26, 43, 62, 80, 82, 106, 108, 110, 122, 129–130, 160, 199, 209
archaeology 5, 35, 45, 94–95, 130, 150, 157, 165, 196, 210

architects 13, 82, 165, 202
architectural evaluation 81
architectural techniques 41
architectural traits 184
architecture 5, 26, 31, 63, 84–85,
    146, 167, 177, 179, 181, 184,
    187, 190, 204–205, 245–246
  contemplative 256
  megalithic 4, 74
  sacred 189, 241, 247
art historian 13
attention deficit hyperactivity
    disorder (ADHD) 161
AUB, see Absolute Unitary Being
audible resonance frequency 83
aural architecture, sensitively
    designed 178
autism 161
awareness 129, 238
  archaeoacoustic 162
  spiritual 187
Aztec sound heritage 14

Babylonian materials 91
Baptistry of St. John adjacent
    to Pisa's Leaning Tower
    181–182, 184
barley 62, 64
basilicas 241, 246–247
bass singing voice 37
bouncing sound 181
boundary-marking voice training
    251
brain 5, 8, 31, 41, 51–52, 147,
    149–150, 152–154, 161,
    164, 180, 186–189, 192,
    201, 219–221, 234, 236,
    238, 258
  frontal lobe 187–188
  parietal lobe 187–188

brain activity 40, 147, 149, 152,
    167, 184, 220, 236, 238
brain areas 192
brain cells 5
brain circuitry 151
brain functions 236
brain imaging studies 153
brain scan studies 187
brain's reward system 150
brainstem 147
brainwave synchronization 180,
    258
brainwaves 179
Buddhist caves 245

Çatalhöyük 60, 62, 120, 122–123,
    125
cathedral 237, 246, 251
cattle 62, 64
cave environments 164
cave paintings 52, 205
caves 7–9, 52–53, 81, 84, 99,
    106–107, 122, 124, 133, 164,
    198, 202–205, 218–219, 224,
    237, 241–242, 244–245,
    248–254, 256
  ancient 244
  echoing 122
  natural 97
  painted 9, 43, 98–99, 105, 111,
    195, 198, 200, 202–206
  paleolithic 52, 133
  prehistoric 254
  resonant 252
  revered 81
  ritual use of 99, 106
  sacred 106, 164
  sculpted 245
  side 123
  underground 202

ceilings 24, 26, 30–32, 34, 38, 52, 69, 90, 131, 182, 184
  corbelled 42
  low 111, 201
  tray 30
  tray-stepped 97
  vaulted 96
cerebral blood flow 151
ceremonial environment 153
ceremonies 105–106, 119, 123–124, 129, 164, 216–218
chamber ceiling 133
chanting, echoing ritual 41
chanting tones 259–260
chapels 196, 202–203, 205
children 39, 45, 65, 96, 120, 124, 165, 197, 223, 250, 254
Christian Era 202–203, 205, 207
chronic relapsing brain disorder 151
churches 115, 133, 153, 181–182, 205, 246
cognition 24, 231–236, 238
Coimbra, Fernando 137, 149
communication, spiritual 242–243
complex building techniques 83
concave walls 42
consciousness 41, 78, 120–121, 133, 149, 162, 164, 177, 179, 188, 190, 192, 207, 235, 238, 248, 250, 254, 257–258
  deepest 250
  early 162
  holistic medical 258
consciousness music 254
construction, architectural 246
contemplative spaces 242
convergence of sound and space 177, 179, 181, 185, 187, 189, 191, 193
Cooke, Miriam 49
cortical activation 154–155

cupola 242–244, 248
curved walls 91
cymatic explorations of tones 191
cymatics 178, 190

depression 161, 163, 167, 189, 193, 222, 249
design
  architectural 11
  monumental 100
  mystical 131
Devereux, Paul 11, 39, 127
digeridoo 149
DNA 64, 68–69
dogs 52, 74
dolichocephalic 45
dopamine 150, 152
dopamine release 147, 150
drum 15, 36–38, 108, 112, 152–153, 201, 219
drumming 15, 187
drumming sound 39

eardrum 36, 147
ears 1, 31, 36, 121, 137, 147, 180, 245, 250, 252
earthquakes 221
  monitoring 83
earth's forces 83
earth's revolutions 179
Egyptian temples 202
Elephant Island 244–245
Elephanta Caves (Mumbai) 245
emotional arousal 222
emotional attention 222
emotional experiences 31, 217, 222
emotional responses 188, 223

intense 188
  pleasant 151
emotional seesaw-effects 43, 210
emotional stimuli 222
emotions 35, 104, 149, 151, 212, 214, 220-221, 223
  deep 130
  negative 222
  painful 13
  positive 221
  positive social 221
  spellbinding 224
  stirring 104
enclosure
  megalithic 85
  monumental 57, 73
endoneural world 236
energy 51, 83, 85, 181-182, 244, 248, 250
England 11, 39, 93, 111
enlightenment 186-189, 192, 224
environment 1, 8, 13, 40-41, 44, 52, 130, 177, 181, 186, 231-236, 238, 255, 257
environmental pollutants 179
ethnographic societies 214, 217
  modern 211
ethnomusicology 5, 241
Europe 40, 64-65, 69, 113, 115
Europe's megalithic buildings 46
excavations 21, 26, 74, 78, 95, 107-108, 110, 122, 130, 141, 256
ExoBuilding experiment 179
exosomatic world 236
experiential mini-museum 167

farming 4, 23, 61, 63-64
  dairy 65

fear 159, 210, 212, 214, 221-222, 224-227, 237
fear and amazement 43, 209-226
fear-induction 225
fear-then-relief 209, 223-224
female high voices 111, 200
female voice 247
frequencies 15, 36-38, 40-41, 65, 99, 111-112, 178-181, 186, 190, 200-202, 219, 238, 245
  common 111
  cosmological 201
  high 42, 111, 200
  low 42, 201
  singular 179
  universal 201
functional magnetic resonance imaging 149-150, 221

geneticists 164
genetics 5, 62-65, 67, 69, 151, 164, 210
Georgian polyphonic singing 253
globigerina 97, 99
Göbekli Tepe (Turkey) 4, 21, 23, 25, 46, 57-59, 61-65, 69-70, 73-74, 78, 80, 85-86, 89, 91, 94-96, 98-99, 108, 110-111, 115, 119-120, 122-123, 126, 137, 140-141, 146
  artisans 74
  excavation site 55-56
Gozo 17, 21-23, 26, 124-125
  megalithic temples 22
  megalithic temples of 22
Grand Song of the Dong ethnic group, Guizhou Province, China 109
Great Pyramid 257, 259-260
  King's Chamber 259
Greek temples 201-202

Hagar Qim site 82
Hal Saflieni Hypogeum (Malta)
	11–12, 24–32, 34–35,
	37–46, 90, 96, 99, 105,
	110–112, 115, 120–121,
	123–124, 128, 131–133,
	137, 149, 161–162, 184–187,
	190, 195–200, 202–207,
	209–210, 218–222, 225–227,
	243
  acoustics of 37, 184
  megalithic interiors 90
Hal Saflieni Project 21–46
hard reflective surfaces 96–97
harmonic sounds 111, 200, 202
healing voice 247
heart rate 150–151, 179
Hindu caves 245
human body 108, 178–179, 254
human consciousness 115, 233, 250
human cultures 217, 224, 235
human development 2, 4, 9, 53, 81, 130, 167
human experience 5, 13, 181, 190
human mind/body/brain 233
human voice 108, 162, 242–243, 245, 247–248, 250
  live 162
hyper-acoustic places 43
hyper-acoustics 218–219
hyper-resonant 190
Hypogeum, the, *see* Hal Saflieni Hypogeum
Hypogeum paintings 205

Indo-European languages 65, 68
infrasound 83–84
initiation-rites 210, 216, 225–227
inner-body massage 251

inner voice 247–249, 251–252, 255
Ireland 11, 39, 46, 62, 70, 111

Karahantepe 6, 45, 78, 80, 99
King Solomon's Temple (Jerusalem) 97
knowledge
  ancient 12
  audiologic 46
  primitive holistic 160
  scientific 10
Kundt's Tube 126–127

Late Stone Age 5
Lespugue 203–204
limestone 7, 97–99
  natural 23
  pale ivory 17
  soft 86
limestone caves 7, 47
limestone pillars 115
limestone seats 98
low acoustics 203
low resonance 202
Lubman, David 157

male voice 32, 111, 199–200, 207, 219
  low 36, 38
Malta 3, 11–12, 14–17, 21–29, 38, 41, 43, 46, 62–63, 65, 67, 83, 86, 91–92, 95–98, 106, 110, 119, 122–125, 128, 131–133, 135–136, 138, 149, 184, 195–198, 205, 209–210, 218–219, 226–227, 242–243

buildings 46
Grand Harbour 25
islands 25, 121, 124
megalithic architecture 202
megalithic monuments 24
megalithic shrines 46
megalithic temples 22, 85
Mnajdra Temple site 129
monuments 86, 94, 99–100
Neolithic building 46
Neolithic ruins 63
Neolithic Temple Culture 137
prehistoric monuments 65
prehistoric temple builders 12
prehistoric Temple Culture 26
sediment cores 23
temple structures 89
temples 21, 69, 74, 82, 85, 88, 94, 96, 99
Maltese limestone 97
material reflectance 190
Medamoud Temple 202
medical research 105
megalithic monuments 21, 23, 61, 69, 100, 147
megalithic sites 39
  ancient 11
megalithic soundboxes 161
megalithic structures 21, 91–93, 154, 259
megalithic temples 23–24, 125
  ancient 97
  freestanding 25
megaliths 4, 62, 73–100
Mesopotamian music 113
Mesopotamian proto-music 113
midbrain 151
Mnajdra South Temple (Malta) 3, 16, 91, 128
monumental architecture 80
  symmetrical 124

monuments 4–5, 9, 13, 15, 47, 57, 62, 85, 89, 94–95, 98–99, 110, 119–120, 133, 198–199
  ancient 15, 101
  chambered 137
  underground 196
  visible standing 21
Mother Earth 133, 149, 207, 249, 251–252
  bones of 81
mountains 70, 99, 108
Mumbai 244–245
music 4–5, 13, 35, 37, 41, 103–115, 130, 147, 150–155, 161, 164–165, 180, 187, 195, 197, 211–214, 217, 225, 232, 234–237, 241–243, 248, 250, 253–255, 258–259
  ancient 197
  early 246
  new-age 257
  non-hunter-gatherer 213
  subject-selected 151
music listening 150
music memories 161
music perception 149
music processing 164
music therapy 114, 180
musical hum 117
musical instruments 35, 104, 107, 198, 201
  ancient 164
  first stringed 108
musical meaning, primitive
  elements of 250, 253–255
musical notation 32
musical training 114

Na'ii'ees 217
Native Americans 74, 120

Neolithic 26, 47, 58, 63, 69, 119, 122–124, 129–131, 137, 207, 227
  pre-pottery 57
Neolithic age 125
Neolithic architects 202
Neolithic cultures 62, 125
Neolithic deposits 124
Neolithic era 2
Neolithic farmers 69
Neolithic figurines 65
Neolithic inhabitants 69
Neolithic Kit 62
Neolithic migrations 68
Neolithic period 62, 115, 122
Neolithic period excavations 108
Neolithic populations 149
Neolithic Revolution 14, 46, 58–59, 61
Neolithic settlers 65
Neolithic societies, complex 58
Neolithic society 21
Neolithic Stone Age 11
Neolithic temple spaces 121
Neolithic tribes 196
neurological research 10, 147
neurons 188, 236
neuroscience 5, 147, 149, 151, 153, 155, 164, 177–178, 186
  cognitive 4
  modern 155
New Stone Age 2, 26
Newgrange Passage 127
Newgrange Passage Tomb 32, 40, 127
noise 38, 105–106, 249–250, 254–255

ochre paintings 200, 203
  geometric 205
  red geometric 196

paintings 30, 52, 81, 124, 130–131, 133, 195, 203, 205–206, 233, 245
  fresco 110
  geometric 205
Palaeolithic 146, 203, 205, 242–243, 246, 251
Palaeolithic Era 202–203, 205, 207
Palaeolithic painted caves 195
Pan rituals 107, 164
perception
  aural 250
  deepest prenatal 162
  human spatial 178
  somatosensory 187
pitch 8, 25, 31, 37, 39, 41, 111, 153–154, 161, 186, 199–202, 219–220, 222, 244–245, 253
portal post stones 90
portals 26, 89, 95, 184
  megalithic post-and-lintel 94
porthole-stones 95
positron emission tomography 150–151, 155, 221
pottery 21, 57, 163
pottery sounds 163
prayers 105, 196, 207, 220, 245, 254
  musical 217
pre-pottery Neolithic site 141
prefrontal cortex 40, 149, 186, 188, 219–220
pregnancy 162, 250
prehistoric acoustic plan 91
prehistoric ancestors 38
prehistoric art 137
prehistoric artwork 131
prehistoric burial site 26
prehistoric colonists 23
prehistoric megalithic monuments 219

prehistoric paintings 149, 198
prehistoric settlement 58
prehistoric temple culture of Malta
   26, 122–123, 138
psycho-acoustics 258
psycho-neurological research 221
pyramids 21, 25, 40, 46, 97, 110,
   154, 259

range
  megalithic 35, 37, 39–40, 111,
     152–153
  vocal 178
religious experiences 115, 153
  intense 115
resonance 37, 39, 99, 105,
   111–112, 161–162, 181, 184,
   195, 199–203, 207, 219, 222,
   241–242, 244–245, 247, 250,
   255–256, 259
resonance frequencies 37, 40, 202
resonant chambers 114
  in voice therapy 255
resonant form 177–192
resonant frequency 37, 184, 186,
   190
reverberation 181
reverbs, deep 257
rhythm 35, 179, 186, 250, 254,
   258
rhythmic patterns 188
rhythmic ritual 115
rhythmic sounds of tapping and
   drumming 187
rites-of-initiation 43, 224–225
ritual 7, 26, 52, 63, 105–107, 109,
   119, 123, 130, 153, 189, 192,
   218, 232, 245
  ancient 5
  contemporary 192

  evocation 107
  religious 35
  secular 189
  ritual worship 39
Romanesque chapels 105, 111,
   195, 200
Romans 17, 25, 97–98, 121

sacred spaces 43, 192, 242,
   245–247, 256
Sapient Paradox 4–5, 7, 9, 61
Sardinia 46, 68, 91, 93, 110, 124,
   205
Schmidt, Klaus 53
Schumann resonance 259
scientific research 161
secondary auditory cortex
   187–188
seesaw-of-emotions 209–210,
   223–227
sensory spaces 177
shamanistic cultures 120
shrines 24, 26, 57, 74, 80, 153
Sicily 17, 23, 46, 91
silence 1, 15, 31, 38–39, 84, 104,
   110, 186, 197, 222, 232, 242,
   248, 255–256, 258–259
singing 7–8, 13, 31, 104–105,
   113–114, 154, 164, 186, 191,
   200, 207, 217, 220, 242–243,
   246–247, 252–255
  archaic polyphonic 68
  ceremonial 39
  folk song 113
  overtone 247
Sleeping Lady, the 137, 196–197
social compliance 226–227
societies
  egalitarian 125
  fishing 212

gatherer   214
hunting-gathering   212
modern   45, 210
pastoral   212
subsistence farming   212
temple-building Neolithic   24
songs   7, 108–109, 152, 238, 245, 247
sonic environment   180, 182
sonic instruments   178
sonic phenomenon   177, 190
souls   26, 35, 109, 114, 133, 189, 205, 244, 249, 255–256
sound
  anthrophonic   231–232, 238
  archaeology of   13–14, 119, 167, 242–243
  bass   15
  dark   221
  musical   40
  mystical   244
  natural   106, 112
  resonant   5, 167
  reverberant   237
  reverberating   137
  sacred   46
  sonorous   107
  spatial   181
  strong   226
  vocal   235
sound-based therapies   193
sound behavior   25, 38, 43, 65, 147, 241–242, 245
sound effects   39, 41, 137, 164
sound healers   13, 40, 257
sound healing   162
sound manipulation   14
sound perception   254
sound quality   30, 57, 80, 90, 246
sound resonance   203, 242–243, 246, 251
sound therapy   161, 195, 201–202, 256

sound waves   32, 36, 38, 42, 112, 147, 182, 184, 186, 255
sounds, fear-provoking   209, 226
space
  acoustic   203, 242, 244
  ancient monumental   105
  ancient ritual   44
  ceremonial   5
  dark   127
  enclosed   97, 115, 129
  hyper-reverberant   190
  immersive   180
  inhabitable   190
  monumental ritual   110
  resonant   153, 198, 237
  resonant cave   233
  reverberating   121
  sensory expanding   187
  subterranean   99
spine   253
spirals   84, 133, 205
  megalithic   133, 205
spiritual feeling   153
spiritual power   245
spirituality   186
startle-response   222
Stone Age   2, 11, 25, 46, 62, 65, 68, 151
Stone Age soundtracks   127
stone basins   63
stone cavities   40
stone chambers, ancient   152
stone circles   21, 57, 73, 80
stone cliffs   106
stone dolmens   94
stone enclosures   57–58
stone lamps   52
stone monuments   4, 14, 46
stone roller balls   82
stone shrines   80
  post Ice-Age   21
stone structures, monumental   81

stone tools 2
stone vats 63
stone walls 15, 52, 74
stones
  carved 133
  corbelled 90
  decorated 86
  hard 96
  lintel 90
  lithophonic 103
  megalithic 85
  pallid 19
  porthole 95
  sacred 126
  sedimentary 99
  smooching 108
supernatural 43, 106, 120, 153, 244, 251
symbolic death 226–227
  fear-inducing 226
symbolic-death-experiences 224

T-shaped stones 141
Tarxien Temple (Malta) 67, 82, 135
Taurus Mountains 99
temple builders of Malta 21
temples
  ancient 14
  ancient storm-god 99
  oldest 73
  semi-subterranean 83
  singing stone 43
  twin Ġgantija 21
thermal baths 241, 243
tombs, prehistoric 124
tones 36, 112, 126, 167, 180, 182, 186–187, 190–191, 258
  artificial electronic 161
  deep 32

echoed 120
electronic 152
lower 162
main 201
resonant 186
traditional singing 241, 249
traditions, megalithic 14, 69
transcendental resonance 149
Trump, David 123
Turkey 6, 9, 21, 23, 25, 39, 53, 62, 73, 83, 92, 95, 99, 137, 175

universe 40, 153, 188, 201, 236–238
Urfa man 137, 140

vibrations 36–37, 46, 83, 111–112, 127, 147, 163, 250, 253–254
  energetic 114
  highest 254
  low frequency 37
  mental 114
  natural infrasound 84
  physical 114, 255
  sensing 250
  slowest 36
  spiritual 256
  tingling 186
vocal cords 35, 104, 108, 114
vocal energies 182
voice box 146
voice-ladder 253
voice of authority 101
voice perception 256
voice scale 251
voice therapists 251
voice therapy 241–256
voices

baritone   111, 161
contemporary   256
female   32, 111, 256
low   30, 39, 152–153
noisy   260
normal   199
strong   200
supernatural   120
well-trained   37
vowels   252–253

wall paintings   197
Western musical scale, modern   111
whistling   107–108, 221
wind instruments   106, 108
worship, ancestor   125

Y-haplogroup   64